Schriftenreihe des Energie-Forschungszentrums Niedersachsen (EFZN)

Band 67

AF457827

Das EFZN ist ein gemeinsames wissenschaftliches Zentrum der Universitäten:

Steady State and Time Dependent Compressed Air Energy Storage Model Validated with Huntorf Operational Data and Investigation of Hydrogen Options for a Sustainable Energy Supply

DOCTORAL THESIS
(DISSERTATION)

to be awarded the degree
Doctor of Engineering (Dr.-Ing.)

submitted by

Friederike KAISER
from Marburg on the River Lahn

approved by the Faculty of Energy and Management

CLAUSTHAL UNIVERSITY OF TECHNOLOGY

Date of oral examination
July 29, 2020

Bibliografische Information der Deutschen Nationalbibliothek
Die Deutsche Nationalbibliothek verzeichnet diese Publikation in der
Deutschen Nationalbibliografie; detaillierte bibliographische Daten sind im Internet
über http://dnb.d-nb.de abrufbar.
1. Aufl. - Göttingen: Cuvillier, 2020
Zugl.: (TU) Clausthal, Univ., Diss., 2020

Dean:	Prof. Dr. rer. nat. habil. Bernd Lehmann
Supervising tutor:	Prof. Dr.-Ing. Roman Weber
Reviewer:	Prof. Dr.-Ing. Hans-Peter Beck
Reviewer:	Dr.-Ing. habil. Marco Mancini

D 104

© CUVILLIER VERLAG, Göttingen 2020
Nonnenstieg 8, 37075 Göttingen
Telefon: 0551-54724-0
Telefax: 0551-54724-21
www.cuvillier.de

Alle Rechte vorbehalten. Ohne ausdrückliche Genehmigung des Verlages ist
es nicht gestattet, das Buch oder Teile daraus auf fotomechanischem Weg
(Fotokopie, Mikrokopie) zu vervielfältigen.
1. Auflage, 2020
Gedruckt auf umweltfreundlichem, säurefreiem Papier aus nachhaltiger Forstwirtschaft.

ISBN 978-3-7369-7341-1
eISBN 978-3-7369-6341-2

Declaration of Authorship

I, Friederike KAISER, declare that this thesis titled, "Steady State and Time Dependent Compressed Air Energy Storage Model Validated with Huntorf Operational Data and Investigation of Hydrogen Options for a Sustainable Energy Supply" and the work presented in it are my own. I confirm that:

- This work was done wholly or mainly while in candidature for a research degree at this University.
- Where any part of this thesis has previously been submitted for a degree or any other qualification at this University or any other institution, this has been clearly stated.
- Where I have consulted the published work of others, this is always clearly attributed.
- Where I have quoted from the work of others, the source is always given. With the exception of such quotations, this thesis is entirely my own work.
- I have acknowledged all main sources of help.
- Where the thesis is based on work done by myself jointly with others, I have made clear exactly what was done by others and what I have contributed myself.

"Wem Mutter Natur ein Gärtchen gibt und Rosen, dem gibt sie auch Raupen und Blattläuse, damit er's verlernt, sich über Kleinigkeiten zu entrüsten."

Wilhelm Busch

Abstract

Wind power and photovoltaic energy play a significant role in sustainable energy systems. However, these two renewable energy sources do not generate electrical energy on demand and are subject to natural fluctuations. Thus, the need for compensatory measures arises. Compressed air energy storage power plants (CAES) are a possible solution to providing negative and positive control energy in the electric grid. However, in contrast to other energy storage devices such as pumped hydro energy storage or batteries, the storage medium *compressed air* hardly contains any energy (or more precisely: enthalpy). Yet, compressed air storage allows the operation of highly efficient gas turbines, which are not only particularly fast available but also achieve better efficiency than combined cycle power plants used today, as illustrated by the example of the modern gas and steam power plant Irsching with $\eta_{tc} = 60$ % from 2011 compared to the 20 years older McIntosh CAES with $\eta_{tc} = 82.4$ %.

In this thesis, the calculation methods for the thermodynamics of the CAES process are presented and validated by measured data from the operations of the CAES power plant Huntorf. Both the steady state and the dynamic (time-dependent) analyses of the process take place. The characteristic value *efficiency* is discussed in detail, since numerous different interpretations for CAES exist in the literature. A new calculation method for the electric energy storage efficiency is presented, and a method for the calculation of an economically equivalent electricity storage efficiency is developed. Consideration is given to the transformation of the CAES process into a hydrogen-driven and, thus, greenhouse gas-free process. Finally, a model CAES system is tested in a 100 % renewable model environment.

Consequently, it can be stated that in the steady-state thermodynamic calculation in particular, the consideration of realistic isentropic efficiencies of compressors and turbines is essential to correctly estimate the characteristic values of the process. Furthermore, a steady-state view should always be accompanied by dynamic considerations, since some process characteristics are always time-dependent. The simulation shows that by mapping transient operating conditions, the overall efficiency of the system must be corrected downwards. Nevertheless, in the model environment of a 100 % renewable energy system, it has been shown that a CAES is a useful addition that can provide long-term energy storage.

Zusammenfassung

In einer nachhaltigen Energiewirtschaft nehmen Windkraft und Photovoltaik eine wichtige Stellung ein. Diese beiden erneuerbaren Energiequellen erzeugen elektrische Energie jedoch nicht bedarfsorientiert, sondern unterliegen natürlichen Schwankungen. Somit steigt der Bedarf an Ausgleichsmaßnahmen. Druckluftspeicherkraftwerke (CAES, vom Englischen "Compressed Air Energy Storage") sind eine Möglichkeit negative und positive Regelenergie im Netz bereit zu stellen. Im Gegensatz zu klassischen elektrischen Energiespeichern, wie Pumpspeichern und Batterien, enthält das Speichermedium Druckluft jedoch kaum Energie. Dennoch ermöglicht die Speicherung von Druckluft den Betrieb von höchst effizienten Gasturbinen, die nicht nur besonders schnell verfügbar sind, sondern auch bereits vor über 25 Jahren eine bessere Effizienz erreicht hatten, als modernste heutige Kombikraftwerke.

In dieser Arbeit werden Berechnungsmethoden für die Thermodynamik des CAES Prozesses vorgestellt und mittels Messdaten aus dem Betrieb des CAES Kraftwerkes Huntorf validiert. Dabei erfolgt sowohl die stationäre als auch dynamische Analyse des Prozesses. Die charakteristische Größe *Wirkungsgrad* wird ausführlich diskutiert, da in der Literatur zahlreiche unterschiedliche Interpretationen existieren. Eine neue Berechnungsmethode für den elektrischen Speicherwirkungsgrad wird vorgestellt sowie eine Methode zur Berechnung eines ökonomisch äquivalenten Stromspeicherwirkungsgrades entwickelt. Es werden Überlegungen zur Umgestaltung des CAES Prozesses in einen Wasserstoff-getriebenen und somit Treibhausgas-freien Prozess angestellt. Schließlich wird ein Beispiel CAES System in einer 100 % erneuerbaren Modell Umgebung getestet.

Im Ergebnis ist festzustellen, dass bei der stationären thermodynamischen Berechnung insbesondere die Berücksichtigung von realistischen isentropen Wirkungsgraden von Kompressor und Turbine wesentlich ist um die charakteristischen Werte des Prozesses korrekt einzuschätzen. Weiterhin sollte eine stationäre Betrachtung immer auch von dynamischen Betrachtungen begleitet werden, da einige Prozesscharakteristiken stets zeitabhängig sind. In der Simulation zeigt sich, dass durch die Abbildung von instationären Betriebszuständen, der Gesamtwirkungsgrad der Anlage nach unten korrigiert werden muss. Dennoch konnte in der Modellumgebung eines 100 % erneuerbaren Energiesystems gezeigt werden, dass ein CAES eine sinnvolle Ergänzung darstellt, um Energie zu speichern.

Acknowledgements

Diese Doktorarbeit entstand in meiner Zeit als wissenschaftliche Mitarbeiterin am Energie-Forschungszentrum Niedersachsen (EFZN) und dem Forschungszentrum Energiespeichertechnologien der Technischen Universität Clausthal. Daher möchte ich zunächst Prof. Wolfgang Busch danken, der mir die Möglichkeit gab mich im Themenbereich der Speicherung elektrischer Energie zu entfalten. Ebenso gilt mein Dank Prof. Hans-Peter Beck, unter dessen Leitung die Arbeit am EFZN große Freude bereitet hat und sehr lehrreich war.

Ganz wesentlicher Dank gebührt meinem Doktorvater Prof. Roman Weber, der die Ausgestaltung des hier bearbeiteten Themenkomplexes maßgeblich beeinflusst hat, indem er seine Expertise und Zeit in die Betreuung investiert hat. An dieser Stelle auch besten Dank an Dr. Marco Mancini, der mich bei verschiedenen Fragestellungen unterstützt hat.

Ein wichtiges Element dieser Arbeit ist die Verwendung von Messdaten aus dem Betrieb des Druckluftspeicherkraftwerks Huntorf, die Dank der freundlichen Unterstützung von Herrn Uwe Krüger möglich war. Für die gute Zusammenarbeit möchte ich mich daher ganz herzlich bedanken.

Schließlich gilt ganz besonderer Dank meiner Familie.

Contents

List of Figures

List of Tables

List of Abbreviations

ACAES	Adiabatic Compressed Air Energy Storage
ADELE	Adiabatic Compressed Air Storage for Electricity Supply
B	Burner
BOP	balance of plant
C	Compressor
CAES	Compressed Air Energy Storage
CAS	Compressed Air Storage
EOS	Equation of State
G	Electricity generator
HP	High Pressure
HPC	High Pressure Compressor
HPT	High Pressure Turbine
ISACOAST	Isobaric Adiabatic Compressed Air Energy Storage
LCV	Lower Calorific Value
LP	Low Pressure
LPC	Low Pressure Compressor
LPT	Low Pressure Turbine
M	Electricity engine
PHES	Pumped Hydro Energy Storage
PV	Photovoltaic
S	Storage place
T	Turbine
TES	Thermal Energy Storage

List of Symbols

Roman Symbol	Name	Unit
A	surface	m^2
ar	air rate	–
c_p	specific heat capacity at constant pressure	J/kgK
c_v	specific heat capacity at constant volume	J/kgK
E	energy	J
ex	exergy	J/kg
h	specific enthalpy	J/kg
hr	heat rate	kWh/kWh
i	state point	-
k	thermal transmittance	W/m^2K
m	mass	kg
M	molar mass	$kmol/kg$
$\dot{m}$	mass flow rate	kg/s
P	power	W
p	pressure	$bar(=10^5 Pa)$
Q	heat	J
q	specific heat	J/kg
$\dot{Q}$	heat flow rate (fuel enthalpy rate)	W
R	gas constant	$J/kmolK$
r	radius	m
s	specific entropy	J/kgK
T	temperature	K
t	time, duration	s
U	inner energy	J
u	specific inner energy	J/kg
V	volume	m^3
v	specific volume	m^3/kg
$\dot{V}$	volume flow rate	m^3/s
W	work (or Wobbe index)	J (or MJ/m^3)
w_t	specific technical work	J/kg
x	space coordinate	m
z	compressibility factor	-
Greek Symbol		
β	compression ratio	-
η	efficiency	-
ρ	density	kg/m^3
ξ	exergetic efficiency	-
λ	air ratio or heat conductivity	- / W/mK

Index	Description
0	standard conditions
ad	adiabatic
c	charging / compressor
cc	compression conversion
d	discharging

el	electrical
G	generator
i	state point
in	initial/inlet conditions
M	motor
out	outlet conditions
ref	reference
rt	round trip
t	turbine
tc	turbine conversion
th	thermal
thermod.	thermodynamic
∞	environment conditions

Chapter 1

Storage of Renewable Energy – Introduction and Scope

1.1 Motivation and Objectives

Transition from fossil to renewable energy. The German "Energiewende" policy encourages the construction of power plants for renewable energy such as wind farms, photovoltaic (PV), or biomass power plants. The political goal is to reach a share of at least 80 % renewables in the electrical energy supply by 2050 [Bundesministerium für Wirtschaft und Technologie and Bundesministerium für Umwelt, Naturschutz und Reaktorsicherheit, 2012]. The ambitious goal to reduce green house gas emissions comes along with major technical [Scholz et al., 2014] and regulatory/economic challenges [Dehmer, 2013; Narbel and Hansen, 2014]. Hence, it is possible that the politically set timetable is a too short period and that the project of restructuring the overall energy system might take a century or so. However, a general feasibility of 100 % renewable energy supply of all energy sectors (electricity, industry, mobility and heating) is confirmed by regional back-casting scenarios [Faulstich et al., 2016b]. Yet, wind and solar power, which are the two fastest growing renewables, generate electricity naturally intermittent, i.e. not demand-oriented. Thus, new strategies to balance electricity generation and demand are required. In the last decades many solutions, such as increasingly flexible generation capacities, demand side management, sectoral interlinking, and various energy storage options have been suggested. The energy system has to meet many different requirements. It is found that a multitude of different new technologies will have to be used, including several energy storage options for different applications, such as grid stability (short term storage or ancillary electric grid services), intra-day or weekly energy balancing (intermediate storage), and seasonal energy balancing (long term storage) [Beck et al., 2013; Sørensen, 2007; Schulz, 2015].

Required storage capacities. The future storage capacities needed in Germany and other countries depend on a large number of variables and are unknown. Estimates cover a wide range of possible values; e.g. Pape et al. [2014] estimate that in the German grid no electrical energy storage is required for a renewable share of up to 90 % as long as other flexibility options, such as flexible gas turbines and demand side management, are used. Weiss [2013] estimates that an electrical energy storage capacity of 1 to 1.5 TWh is needed in an 80 % renewable energy system and emphasizes that the ratio of wind to PV capacities is a decisive factor for storage assumptions. Scholz et al. [2014] estimate for a 100 % renewable system in Germany a required energy storage capacity of 191 TWh. To date such large energy storage capacities are feasible with chemical energy storage only [Scholz et al., 2014]. Based on a back casting scenario developed in [Faulstich et al., 2016b], Faulstich et al. [2016a] investigate the trade off between long-term chemical energy storage (with relatively low round trip efficiencies) and intermediate-term energy storage with higher round trip efficiencies based on a simulation of the overall energy system of Lower Saxony, a region in northern Germany. They [Faulstich et al., 2016a] estimate that a 100 % renewable energy supply of Lower Saxony requires 33.5 TWh of long-term energy storage capacity in the form of hydrogen storage in addition to 146 GWh intermediate-term energy storage capacity of the pumped hydro energy storage-type or similar. The energy needs of Lower Saxony correspond to approximately 1/10 of the German ones. Hence, it can be deduced that the estimates in [Faulstich et al., 2016a] correspond to a required storage capacity for a 100 % renewable German energy system (comprising electricity, industry, heating and mobility sectors) of 335 TWh long-term and 1.5 TWh intermediate-term storage. Faulstich et al. [2016a] state that for a lower share of renewables (approximately 80 to 90 %) the required storage capacity for Germany is drastically lower of around 11 TWh.

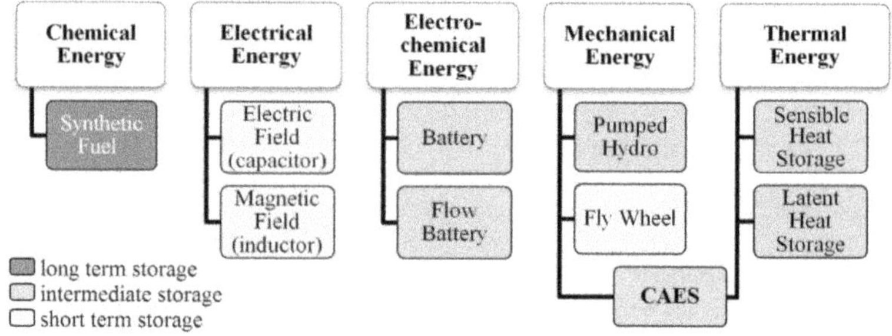

FIGURE 1.1: Electrical energy storage options sorted by form of energy in the storage state

Electrical energy storage. Electricity cannot be stored. However, electric energy can be transformed into storable energy forms such as mechanical energies (kinetic or potential energy), thermal energy, chemical energy, and others. A general overview of electrical energy storage systems sorted by energy form in the storage state is given in Figure 1.1. An example for chemical energy storage is the hydrogen path: Electrical energy is used to produce hydrogen via water electrolysis as a synthetic fuel. Hydrogen can then be stored in its gaseous or liquid form and contains chemical energy. When a deficit in the electric grid occurs, hydrogen can be turned into electric energy via fuel cells or via combustion turbines using oxygen from the air. This hydrogen path allows for large storage capacities and long storage periods due to the stability and storability of fuels such as hydrogen, synthetic natural gas, or Fischer-Tropsch products. However, conversion efficiency is comparatively low. On the other hand, the electrical energy storage in electric fields of capacitors or the magnetic field of inductors has very high conversion efficiencies but storage duration is limited due to high self discharging rates. Thus, together with flywheels which belong to the category of mechanical energy storage systems, these technologies are only applicable to short term storage in the order of milli seconds to minutes. In the section of intermediate term energy storage several technologies are available. Pumped Hydro Energy Storage (PHES) accounts for 99 % of the installed power and capacity in the electric grid [Rehman, Al-Hadhrami, and Alam, 2015; Dötsch, Kanngießer, and Wolf, 2009]. It is in use as bulk energy storage for intra day load shifting for around 100 years. The electrical round trip efficiency is today around 80 %. Compressed Air Energy Storage (CAES) is another option to store energy for intermediate term (storage durations of hours or days). It is often listed in the category of mechanical energy storage due to its seemingly similarities with PHES (e.g. comparable machinery, such as pump or compressor and water or gas turbine, respectively, and similar power and capacity ranges). However, energy flows, efficiencies, and storage mechanisms are different because of the compressibility of the working fluid *air* compared to the incompressible *water* in PHES. Thus, thermodynamic considerations and thermal cycling within the CAES process are also crucial. This makes CAES a thermo-mechanical system. The fact that enthalpy of compressed air at ambient air temperatures is close to zero emphasizes the thermal energy storage characteristics. Without thermal energy storage external heat sources such as natural gas have to be used, turning the overall concept into a combined generation and storage unit.

Compressed Air Energy Storage. The first fuel-driven CAES plant was successfully commissioned in Huntorf in 1978. Today, there are only two commercial plants worldwide - in Huntorf (Germany) and in McIntosh (USA). Due to the switch from fossil to renewable energies CAES gained interest over the last few years and proliferation of publications can be observed; Several pilot plants of next generation CAES systems are recently being tested; Some economic forecasts expect a "dramatic growth" of "more than 11 GW of CAES" until 2023 [Martin, 2013-08-19].

Objectives. In this thesis forthcoming CAES concepts are examined based on steady state calculation methods validated with data from the reference CAES plant Huntorf. Time-dependent aspects of the processes are calculated. Hydrogen as a fuel option is discussed. The calculation methods are validated with a unique set of measured operational data from the reference CAES plant in Huntorf. A hydrogen-fueled CAES model system is tested in a 100 % renewable energy system simulation.

1.2 Previous Publications

In this section previously published literature is summarized sorted by three key topics: CAES in general, time dependent CAES, and hydrogen options for CAES.

1.2.1 Design and Steady State Thermodynamics of CAES

[1] The first patent of CAES dates back to the 1940s [Gay, 1948]. The first plant went in operation in 1978 in Huntorf, Germany [Quast and Crotogino, 1979; Brown Boveri & Cie, 1980; Hoffeins, Romeyke, and Sütterlin, 1980; Quast, 1981; Brown Boveri & Cie, 1986; Crotogino, Mohmeyer, and Scharf, 2001-04-15; Crotogino, 2003]. In the 70s and 80s a research program was conducted in the U.S. to examine underground storage options, in which the concept of CAES was investigated and a large number of potential sites for CAES was identified. Reports of this 'Underground Energy Storage Program' prepared by the Pacific Northwest Laboratory for the U.S. Department of Energy are given in [Drost, Zaloudek, and Loscutoff, 1980; Hendrickson,

[1] This subsection has previously been published in [Kaiser, Weber, and Krüger, 2018].

1981; Hobson et al., 1981; Kannberg, 1981; Wiles and McCann, 1981; Allen, Doherty, and Fossum, 1982; Allen, Doherty, and Thoms, 1982; Reilly and Schainker, 1982; Wiles, 1982; Zaloudek and Reilly, 1982; Fort, 1982; Fort, 1983; Beckwith & Associates, 1983; Erikson, 1983; Hostetler, Childs, and Phillips, 1983] and summarized among others in the annual report [Kannberg, 1983]. Fort [1982] presents the program 'CAESCAP' that was developed by Pacific Northwest National Laboratories (U.S.) in the early 80s and shows methods comparable to those presented in Chapter 3 of this thesis [Fort, 1982; Fort, 1983]. The program serves to evaluate different CAES plant configurations in steady-state conditions including adiabatic CAES (ACAES). Subsequent to these theoretical studies, in the early 90s a CAES plant was commissioned in McIntosh, Alabama, which remains the only operated underground energy storage facility in the U.S. [Goodson, 1993; Pollak, 1994]. Osterle [1991] presents a thermodynamic analysis of a basic CAES configuration based on Clapeyron equation of state (EOS). Compression and expansion are assumed to be irreversible and exergy is used to define a process efficiency.

Seventeen years later, in the context of renewable energy new studies on adiabatic CAES (ACAES) are elaborated. Grazzini and Milazzo [2008] consider various design criteria for ACAES such as the storage pressure and the number of compression or expansion stages. The considerations are based on Clapeyron equation of state (EOS) and reversible thermodynamics. For ACAES a round trip storage efficiency of 72 % has been found [Grazzini and Milazzo, 2008]. In 2011 a more detailed study of ACAES was presented by Wolf [2011] discussing several EOS that might be suitable for CAES calculations. Wolf [2011] questioned the applicability of ideal gas law models for their insufficient representation of the air storage state in which a high pressure together with a low temperature occur. Furthermore, heat storage optimization was a focus in his study.

Hartmann et al. [2012a] analyse different ACAES concepts using again ideal gas EOS and considering reversible and irreversible ACAES with a variation of the number of compression stages (1 to 3 stages). Some dynamic elements are added, since the effect of rising cavern pressure on compression side during the charging process is taken into account. This effect is neglected during discharging, where no throttle is modeled but the turbine power remains constant over a wide pressure range. The main conclusion of Hartmann et al. [2012a] is that the literature value of 70 % ACAES efficiency applies only to reversible models and that 60 % figure seems to be a suitable estimate for an actual plant. Kim et al. [2012] present a thermodynamic analysis of several CAES concepts using both energy and exergy. The calculations are again based on ideal gas law and take irreversibility into account but neglect pressure variations inside the reservoir by evoking isobaric storage solutions. The steady-state calculations [Kim et al., 2012] consider a diabatic plant that resembles the McIntosh plant. Furthermore, isothermal configuration, tri-generation micro-CAES with storage, heating and cooling, micro-ACAES, and isobaric CAES plus a pumped hydro energy storage-hybrid system are considered. Kim et al. [2012] introduce a new CAES storage efficiency based on exergy and conclude that for CAES a figure of around 70 % is applicable. Less optimistically, Pickard, Hansing, and Shen [2009], using ideal gas law, irreversibility, and exergy calculations too, estimate that ACAES attains a cycle efficiency of roughly 50 % to 60 % only. Nielsen [2013] gives an overview of several CAES configurations and describes in detail several variations of an isobaric plant ('ISACOAST-CC'). Furthermore, heat transfer processes inside a salt cavern are considered. The calculations include a dynamic simulation that takes into account a part load operation. Zhao, Wang, and Dai [2015] present a steady-state calculation of CAES combined with a Kalina process where Clapeyron EOS is used. Several parameter variations are presented to show the effect of changing ambient conditions or turbine inlet temperatures. Again exergy is used. Safaei Mohamadabadi [2015] compares several CAES concepts including the conventional CAES, a cogeneration CAES, adiabatic, and hydrogen fired CAES. Although thermodynamic irreversibilities are accounted for, Clapeyron EOS is used and specific heats are taken as constant. The concept of exergy is used to asses efficiencies. An economic assessment is the main focus.

Budt et al. [2016a] present both an overview of CAES history and recent developments. The publication includes considerations on exergy, efficiencies, and fluid properties. Huntorf and McIntosh plants are described using generally available (literature) process data and several advanced CAES systems are elaborated upon in more detail. Castellani et al. [2015] present experimental data on CAES with considerations on heat storage options based on phase changes. Additionally, Tessier et al. [2016] provide theoretical considerations for such an ACAES. Mazloum, Sayah, and Nemer [2016] present a steady-state and dynamic calculation of ACAES resulting in a round trip efficiency of 66 % based on irreversible thermodynamics. Briola et al. [2016] present thermodynamic calculations for the Huntorf CAES plant (original 290 MW configuration before the retrofit to 310 MW) with air treated as real gas and irreversible thermodynamics. The main focus is on dynamic plant behavior based on characteristic curves of the turbo machinery. Even though Huntorf and McIntosh plants play a key role in the above cited literature, thermodynamic insights are often limited to the citation of (generally available) literature values of the process parameters. Only when it comes to dynamic aspects, such as the transient behavior of the cavern the existing plants are considered more thoroughly. Thermodynamic data of the Huntorf air storage cavern was originally presented in [Hoffeins, Romeyke, and Sütterlin, 1980; Quast, 1981] and has been analyzed in works of Raju and Khaitan [Raju and Khaitan, 2012; Khaitan and Raju, 2012; Khaitan and Raju, 2013], Kushnir et al. [Kushnir, Ullmann, and Dayan, 2012b; Kushnir, Dayan, and Ullmann, 2012] and Xia et al. [2015]. Recently, Briola et al. presented dynamics of turbo machinery in more detail [Briola et al., 2016]. Nakhamkin et al. [1989] discuss the transient thermodynamics of McIntosh's cavern.

Conclusion. The corollary of the above literature review on thermodynamics of CAES is that no detailed thermodynamic steady-state analyses has been carried out in which the gas is treated as real and a consistent method validated with measured operational data is used to handle process irreversibilities for both, the existing CAES plants and the forthcoming (conceptual) CAES designs.

1.2.2 Time Dependent Thermodynamics of CAES

CAES and its transient aspects are subject of research and development since the 1970s. Recently, the combined operation of CAES together with renewables is of major interest, but also thermodynamics of the air storage cavern, the heat storage as well as part-load behavior have been addressed in a number of projects and scientific papers. CAES related subjects that require time dependent calculation are:

- Combined operation of CAES with renewables,
- Thermodynamics of compressed air storage cavern (CAS),
- Thermal energy storage (TES) for ACAES,

- Part load, start-up and run down procedures.

Combined operation of CAES with renewables. Due to the unsteady nature of the renewable energy sources wind and solar power, time-dependent analysis of energy systems becomes ever more important. There are several publications dealing with combined operation of **wind power and CAES** [Greenblatt et al., 2007; Ibrahim et al., 2007; Ibrahim et al., 2012; Wolf, 2011; Fertig and Apt, 2011; Mason and Archer, 2012; Mauch, Carvalho, and Apt, 2012; Madlener and Latz, 2013; Maton, Zhao, and Brouwer, 2013; Gu et al., 2013; Yang et al., 2014b; Zhao et al., 2015; Saadat, Shirazi, and Li, 2015; Ramadan et al., 2015; Bosio and Verda, 2015], **solar power and CAES** [Arabkoohsar et al., 2015], **renewables in general and CAES** [Grazzini and Milazzo, 2008; Garvey, 2012] (renewable produce directly compressed air) [Marano, Rizzo, and Tiano, 2012; Berrada and Loudiyi, 2016], CAES on different energy markets [Lund and Salgi, 2009] (spot market), [Drury, Denholm, and Sioshansi, 2011] (energy and reserve market), or, in more general terms, **energy storage with renewables** e.g. [Kondoh et al., 2000; Sundararagavan and Baker, 2012; Weiss et al., 2016]. These studies often aim at an economic optimization of the combined system using a technical plant model that is reduced to a set of characteristic values such as cost, heat rate, and full load operation duration. Some of these characteristic values are presented in the Table 1.1. The CAES heat rate is commonly set to 1.17 kWh/kWh, which corresponds to the $hr_2 = Q_{fuel}/W_{el,turb}$ of the McIntosh CAES plant (see Chapter 3). An electrical turnaround ratio of around 150 % is often used which indirectly includes fuel contribution and corresponds to the formulation $\eta_{rt1} = W_{el,turb}/W_{el,comp}$ (Eq. 2.11 in Chapter 2). In such a ratio the contribution of fuel energy (Q) is not considered which leads to 'efficiency' values > 1. For adiabatic CAES a round trip efficiency of around 70 % is commonly used. In addition, Mason and Archer [2012] use load dependent heat rates in their calculations and are, thus, one of the advanced economic studies. However, despite this seemingly consistent use of η_{rt1} and hr_2 the origin of these values is handled in different manners: When splitting the round trip efficiency η_{rt1} into two efficiencies of charging (compression) and discharging (expansion in turbines) the estimates are quite different (see Table 1.1) which may lead to a significantly different evaluation of the overall process, e.g. Foley and Díaz Lobera [2013] base their considerations on a fuel-driven CAES plant with a realistic heat rate combined with the round trip efficiency of an adiabatic plant. In consequence, resulting considerations must underestimate the potential of CAES by far. Due to these unambiguous treatment of CAES characteristics in literature it is useful to examine CAES efficiency values in detail (see Chapter 2) based on fundamentals of thermodynamics.

Thermodynamics of air storage caverns. Compressed air can be stored in all kinds of pressure vessels [Budt et al., 2016b]. Underground salt cavern have proven their reliability for large scale CAES facilities in Huntorf and McIntosh. There are several articles dealing with the time dependent behavior of air storage. Langham [1965] is the first publication describing simulation of the dynamics of air storage for CAES. Langham [1965] describes a system where compressed air is cooled and stored inside excavated rock tunnels assuming air leakage through fissures and faults. The calculations assume complete mixing of the air; Specific heat capacity is constant. Langham [1965] concludes that air leakage has a major impact on the overall storage set-up. The first CAES plant Huntorf (Germany) was commissioned in 1978 using a salt cavern for air storage. It was found that air leakage of salt caverns is negligible. Some operational data is published in the form of diagrams [Quast and

TABLE 1.1: Characteristic values of CAES used in different studies

Efficiencies charge ·	discharge =	η_{rt1}	hr_2 $\frac{kWh_{th}}{kWh_{el}}$	Reference
		1.50	1.17	Greenblatt et al. [2007]
		1.35	1.17	Mauch, Carvalho, and Apt [2012]
0.80	1.93	1.54	1.22	Mason and Archer [2012]
0.70	2.00	1.40	1.17	Gu et al. [2013]
0.60	2.49	1.49	1.17	Madlener and Latz [2013]
0.75	0.82	0.62		Maton, Zhao, and Brouwer [2013]
0.80	0.90	0.72	1.20	Foley and Díaz Lobera [2013]
adiabatic:		0.68		Wolf [2011]
		0.68		Kaldemeyer, Boysen, and Tuschy [2016]
		0.80		Liu, Woo, and Zarnikau [2017]
McIntosh:		1.36	1.2	Kaiser, Weber, and Krüger [2018]
Huntorf:		1.19	1.7	Kaiser, Weber, and Krüger [2018]

Crotogino, 1979; Hoffeins, Romeyke, and Sütterlin, 1980; Quast, 1981; Crotogino, Mohmeyer, and Scharf, 2001-04-15]. This operational data was used several times to validate calculations of CAES cavern thermodynamics [Raju and Khaitan, 2012; Kushnir, Ullmann, and Dayan, 2012a; Kushnir, Ullmann, and Dayan, 2012b; Kushnir, Dayan, and Ullmann, 2012; Quast and Crotogino, 1979; Khaitan and Raju, 2013; Xia et al., 2015; Zhao, Wang, and Dai, 2015; Marano, Rizzo, and Tiano, 2012; Maton, Zhao, and Brouwer, 2013; Hartmann et al., 2012a]. Raju and Khaitan [2012] calculate the cavern behavior based on the Huntorf values presented in [Brown Boveri & Cie, 1980] and [Crotogino, Mohmeyer, and Scharf, 2001-04-15] which are mainly originally published earlier by Quast and Crotogino [1979]. Almost all later publications refer to [Crotogino, Mohmeyer, and Scharf, 2001-04-15] and [Raju and Khaitan, 2012], whereas original data from Hoffeins, Romeyke, and Sütterlin [1980] and Quast [1981] seem to be rather untouched in recent literature.

Another approach is presented by Schwoeppe, Gose, and Scholz [2008] who investigate a CAS model in order to estimate temperature and pressure values for geo-mechanical stability considerations in the context of an offshore CAES system in Germany.

Besides afore mentioned Huntorf data Osterle [1991] and Marano, Rizzo, and Tiano [2012] refer to McIntosh data for validation purposes. The working group of Marano et al. focuses on hybrid models consisting of a combination of CAES and wind and/or PV [Arsie et al., April 5-7, 2005] [Marano, Rizzo, and Tiano, 2012]. A few other dynamic calculations of CAES do not contain validation data such as [Garvey, 2012; Nielsen, 2013; Zhang et al., 2013]. Tada et al. [1998] show a numerical analysis of the temperature distribution inside the air storage cavern during charging and discharging.

Furthermore, the thermodynamic behavior of gas storage caverns has been extensively studied in the context of natural gas storage and sophisticated tools for cavern design exist, see Table 1.2. These programs have been adapted to the simulation of compressed air. Thus, user manuals and specialist literature is extensive, e.g. the user's manual "Salt Cavern Thermal Simulator" (SCTS) by RESPEC contains extensive formulas on gas storage thermodynamics [Nieland, 2004]. This study includes an analysis to proof that one single bulk temperature and pressure (such as in [Langham, 1965]) represents the gas storage behavior in an appropriate manner. Numerous research articles are dealing with the thermodynamics of salt caverns in the context of natural gas storage and often in the context of rock stability or production rates such as [Hagoort, 1994; Berest and Brouard, 2003; Lux, 2009; Dresen, 2010; Leuger and

Beutel, 2012; Rutqvist et al., 2012a; Rutqvist et al., 2012b; Kruck, Zander-Schiebenhöfer, and Johansen, 2013; Park et al., 2016].

TABLE 1.2: Simulation software of gas thermodynamics in salt caverns

Software Name	Company	Country
SCTS Salt Cavern Thermal Simulator	RESPEC	USA
COS Cavern Operation Simulator	TRANSITION	Poland
GUSTS v2	GEOSTOCK	France
n.a.	KBB	Germany

Thermal energy storage (TES) for ACAES. In the existing CAES plants approximately 95 % of the electric energy taken from the grid is dissipated as heat losses during compression of air (see Chapter 3 and [Kaiser, Weber, and Krüger, 2018]). Thus, recent literature is often focusing on adiabatic CAES (ACAES) solutions that avoid these losses by using a thermal energy storage (TES) unit and enables storage and re-use of compression heat. Yet, the temperatures within the TES fall or rise during charging and discharging, respectively, making the heat transfer inherently time dependent. Possible technical solutions to achieve satisfactory efficiency values have been largely explored. In the 1970s and 1980s a US research program addresses ACAES technical and economic issues [Kreid, 1976; Kreid, 1977; Drost, Zaloudek, and Loscutoff, 1980; Hobson et al., 1981]. Kreid [1976] confirms the general feasibility of ACAES but estimates that it will not be competitive with fuel-driven CAES. Kreid [1977] investigates several TES concepts such as a hybrid CAES system with TES and fuel-firing or the use of an aquifer as TES. However, conventional recuperator systems showed to be the most economic option. Thus, Kreid and McKinnon [1978] present a detailed thermodynamic analysis of ACAES and a hybrid CAES based on irreversible thermodynamics with extensive parametric studies on plant performance parameters, such as efficiency, system heat rate, and turbine heat rate. Drost, Zaloudek, and Loscutoff [1980] investigate six compressed air energy storage systems with an aquifer as thermal storage and assess economics. Hobson et al. [1981] review direct and indirect contact sensible heat storage, latent heat storage, and thermo-chemical energy storage for CAES. Hobson et al. [1981] show the effects of cyclic thermal storage and how the overall storage temperature slightly rises during the first cycles which entails a slight decrease of the overall efficiency. Similar results are obtained later by Wolf, Berthold, and Dötsch [16.06.2009] and Wolf [2011] who present the concept of excess heat that can occur in TES systems due to irreversibility of compression and expansion as well as humidity of ambient air. In [Wolf, 2011] detailed thermodynamic modeling of the overall ACAES process is presented with a one-tank thermocline TES system including some experimental validation of the model assumptions. However, Hobson et al.'s preferred TES system is a two stage direct contact sensible heat packed bed due to economic reasons (lower cost and commercial readiness) [Hobson et al., 1981]. Thus, various heat storage materials are presented; surface and subsurface storage solutions are discussed and design methods are elaborated in detail. Calculation of the TES is carried out with the "MIT-model" (Hamilton in [United States Department of Energy, Electric Power Research Insitute, and Pacific Northwest Laboratory, 1978-05-15], p.271-307). In 2003 to 2006 the EU-project "Advanced adiabatic compressed air energy storage" (AA-CAES) was initiated by Alstom Power LTD, UK [Bullough et al., 2004; Zunft et al., 2006; Jakiel, Zunft, and Nowi, 2007]. One of the outcomes of this project is that packed bed sensible heat (possibly combined with latent heat) TES are the preferable solution. Several follow-up studies exist [Alstom Power et al., 2007; Dietz, 2008]. Bullough et al. [2004] present some economic aspects and a broad range of technical challenges for ACAES: several turbo machinery options are discussed as well as a variety of TES options such as solid and liquid storage materials as well as phase change storage. Zunft et al. [2006] preferred TES solution was a cylindrical pre-stressed concrete vessel with a volume around $10,000m^3$ filled with solid or solid and phase changing storage materials. They [Zunft et al., 2006] present technical and economic results stating that a 300 MW plant (termed "Central Solution") reaches "thermo economic model efficiencies of more than 70 %, and power related investment costs of less than 800 EUR/kW". Yet, in following projects "ADELE" and "ADELE-ING" [Zunft et al., 2012; Moser et al., 2012; Zunft, 2015] higher investment cost were found: Zunft [2015] presents some of the TES systems in more detail, including regenerator storage (compressed air is in direct contact with a solid storage material), regenerator storage with a secondary loop (gaseous heat transfer medium to cool the compressed air and transfer heat to a solid heat storage material), several stages of liquid heat storage materials (thermal oil and molten salt) and low temperature thermal oil as heat storage (multistage process). Zunft [2015] states that the multi-stage low temperature variant is most promising in terms of risk and economics. The feasibility of pre-stressed concrete vessels has been proven experimentally and several other design aspects have been covered. In conclusion, for such a system a capital cost of 1300 EUR/kW_{el} has then been found [Zunft, 2015]. Thus, a demonstration plant has not been realized due to unresolved economic challenges. Zunft [2015] re-affirms the estimated efficiency of (up to) 70 %, yet, some thermodynamic studies suggest lower values: Hartmann et al. [2012a] analyze different ACAES concepts using ideal gas equation of state. They compare reversible and irreversible calculation methods and consider several ACAES plant layouts. In conclusion they estimate that the overall energy storage efficiency of ACAES is around 60 % which is 10 percent point lower than generally estimated in the European and German CAES projects. Yang et al. [2014a] confirm these results by presenting a parameter variation of heat storage effectiveness and pressure loss for a thermodynamic ACAES system model. Their overall system efficiency is in the range of 52-60 % (when considering pressure losses and irreversibility) [Yang et al., 2014a].

Mei et al. [2015] present analysis and experimental data from an adiabatic system termed "TICC-500" that has been commissioned 2014 in a 420 kW scale at Tsinghua University, Beijing. TES consists of three water tanks at different temperature levels. The energy storage efficiency achieved with TICC-500 system is 41 % [Mei et al., 2015].

The concept of packed bed as TES in the context of CAES is recently revisited by Park et al. [2014], Barbour et al. [2015], and Sciacovelli et al. [2017]. Park et al. [2014] propose a TES system underground in the form of a gravel-filled underground cavern. By comparing it to a conventional above ground thermal storage unit they find lower heat losses due to heating of surrounding rock. The concept of packed bed TES (not in the context of CAES) is also extensively investigated by Beasley and Clark [1984] who investigate several parameter of packed bed TES such as void fraction distribution, thermal wall effects as well as energy losses due to dynamic response. Experimental data of laboratory scaled packed bed TES is known from Meier, Winkler, and Wuillemin [1991], Bauer [2001], Anderson et al. [2015], and Cascetta et al. [Cascetta et al., 2015; Cascetta et al., 2016] who present data and corresponding mathematical models. These mathematical models are briefly presented e.g. by Singh, Saini, and Saini [2009] who compare several calculation methods for predicting thermal performance of packed bed TES systems. Bauer [2001] investigates in more depth several heat transfer mechanisms. Cascetta et al. [Cascetta et al., 2014; Cascetta et al., 2015; Cascetta et al., 2016] elaborated the topic quite extensively developing a 1D numerical model (Matlab-Simulink) and a 2D CFD model (Fluent) of the TES based on their measured data from laboratory scale.

The concept of latent heat energy storage for CAES has been re-investigated by Bullough et al. [2004], Tessier et al. [2016], and Castellani et al. [2015]. Tessier et al. [2016] present heat storage based on a cascade of phase changes. Air is treated as ideal gas; compression and expansion are considered as polytropic; the air storage place is considered as isothermal; efficiencies are based on exergy of enthalpy. The overall storage efficiency is estimated to be 85 %. Castellani et al. [2015] present experimental data of an expansion within a pressure vessel that contains phase changing material to estimate the amount of phase change material needed to attain near-isothermal expansion of air.

Qi [2012] presents a detailed study of two high-temperature heat storage systems for ACAES including some experimental validation. Liu and Wang [Liu and Wang, 2016] carry out an analysis of ACAES with exhaust enthalpy recuperation and estimate the use as a cogeneration unit, while Yao et al. [2016] analyze thermo-economic optima of ACAES (without detailed analysis of TES options).

Part load, start-up and run down procedures. When using CAES as a means for flexible compensation of wind or solar power fluctuations, short operation duration necessitate frequent start-up and run-down procedures as well as flexible power outputs by operation in a part load. Hence, these operation modes have to be handled properly to represent the overall process correctly. Operation characteristics of the Huntorf plant are presented in [Hoffeins, Romeyke, and Sütterlin, 1980]. Hoffeins, Romeyke, and Sütterlin [1980] discuss the commissioning protocols of the Huntorf machinery and describe start-up and part load characteristics of compressors and turbines. More general information is published for conventional gas turbines e.g. by Lechner and Seume [2010] and Marx [2012]. Nielsen [2013] describes start-up procedures of turbo machinery for CAES and takes into account part load behavior by using load dependent isentropic efficiency values corresponding to the manufacturer's characteristic diagrams of compressors and turbines (here: Alstom's GT26). Wolf [2011] (p.124) uses a similar approach by adopting a mass flow rate dependent effective isentropic turbine efficiency in equivalence to typical steam turbine performance characteristic. For compressor calculations Wolf [2011] (p.127) uses an approach based on polytropic efficiency.

Mazloum, Sayah, and Nemer [2016] presented a steady state and dynamic calculation of an adiabatic CAES system where start-up and rundown procedures are taken into account by estimating the effect of thermal inertia of heat exchangers and mechanical inertia of the rotational equipment (compressors and turbines). Processes are considered as polytropic and mechanical losses are taken into account. They conclude that the resulting efficiency of transient calculations is approximately 2 percent points lower than for steady state calculations, which is only valid for operation durations that are very long (more than 10 hours of continuous compression and expansion).

1.2.3 Hydrogen Options for CAES

There are three essential aspects when considering hydrogen for CAES:

- Design of the overall CAES process with hydrogen as fuel [Khaitan and Raju, 2012; Schastlivtsev and Nazarova, 2016];
- Implementation of a hydrogen combustion system (which is not only applicable to CAES but applies to any hydrogen fueled process) [Krüger, 2015-02-19; Gobbato et al., 2011; Lee et al., 2010; Ströhle and Myhrvold, 2007; Chiesa, Lozza, and Mazzocchi, 2005; Juste, 2006];
- Storage of large quantities of hydrogen [Crotogino et al., 2010; Horvath, Donadei, and Schneider, 2016; Donadei and Zander-Schiebenhöfer, 22./23. April 2015; Donadei et al., 2015; Pollok et al., 2015; Böttcher et al., 2017].

Design of hydrogen CAES systems. Khaitan and Raju [2012] analyze the use of a sodium alanate storage system for CAES. Their main objective is examining whether the dynamics of such a storage solution are compatible with the dynamics of a CAES process. Schastlivtsev and Nazarova [2016] present an energy storage gas turbine system that incorporates storage of air, hydrogen (H_2), and oxygen (O_2). In their concept, H_2 is combusted with O_2 to steam. The so generated steam is used to preheat compressed air before expanding in a turbine. Schastlivtsev and Nazarova [2016] consider in their theoretical concept oxy-hydrogen steam generation as an existing technology with further research and development needs, which have been addressed in earlier research in detail by Malyshenko et al. (same working group of the joint Institute for High Temperatures of the Russian Academy of Sciences, Moscow) [Malyshenko et al., 2012]. Mohamadabadi [2014] presents a thermodynamic calculation of CAES with hydrogen combustion compared to adiabatic CAES. Hydrogen is considered in a general manner by considering the energy needed to produce hydrogen in the energy balances of the system. The amount of hydrogen is determined via LCV.

Hydrogen combustion. Combustion of hydrogen (H_2) to partially or fully replace carbon fuels has been investigated extensively. The fundamental reaction mechanisms of hydrogen combustion are investigated by Ströhle and Myhrvold [2007].

Combustion of hydrogen with oxygen (not air) has also been investigated: Experimental data of an oxygen-hydrogen steam generator in the Megawatt power class is presented by Malyshenko et al. [2012]. Based on these experiments and simulation, they [Malyshenko et al., 2012] state that the oxy-hydrogen combustion can be described by 5 combustion zones. Complete combustion of the fuel can only be attained if re-circulation of steam is avoided or appropriate counter measures are taken [Malyshenko et al., 2012]. Yet, the complete combustion was practically not achieved - a minimum hydrogen and oxygen flux of 2vol-% in the exhaust gas has been obtained in experiments [Malyshenko et al., 2012]. One central issue is cooling of the combustion chamber walls which poses several problems [Malyshenko et al., 2012]. A major advantage is the fast start-up time of less than 10 seconds [Malyshenko et al., 2012]. This type of combustion has been investigated earlier by Sternfeld et al. in the context of rocket combustor technology (DLR) [Sternfeld et al., 1995; Sternfeld, 1995]. Recently, this approach is reinvestigated [Krüger, 2015-02-19]. Krueger [Krüger, 2015-02-19] examines hydrogen flame behavior for air and oxygen combustion considering steam injection into the combustion. Krueger concludes that such a setup can be used in existing gas turbines with little constraints. Furthermore, two projects, "Greenest" and "Bluestep", have been initiated at the TU Berlin to demonstrate diluted hydrogen-oxygen combustion under engine conditions.

Gobbato et al. [2011] investigate hydrogen-air combustion in heavy duty gas turbines by a numerical simulation and experimental validation. Their [Gobbato et al., 2011] focus is the prediction of the temperature field of an air-hydrogen combustion in a diffusion flame combustor.

Chiesa, Lozza, and Mazzocchi [2005] investigate the effects of hydrogen combustion in gas turbines and suggest options to handle the occurring effects (change of volume flow rate, NO_x emissions and change of thermophysical properties) by diluting combustion products with nitrogen (N_2) or steam as well as taking additional cooling measures (turbine blades cooling). They conclude that a moderate re-design of gas turbines is required but it is generally feasible. However, with elevated dilution rate efficiency slightly decreases. Juste [2006] investigates injection of small quantities of hydrogen only (4 %) in a hydrocarbon fueled burner to reduce the carbon dioxide (CO_2) emissions. He suggests injecting hydrogen in the primary zone, premixed with the air and, thus, achieves lower NO_x emissions. Lee et al. [2010] examine the combustion of hydrogen (H_2) and carbon monoxide (CO) mixtures stating that high hydrogen content in the gas leads to high NO_x emission probably due to high combustion temperatures.

Hydrogen storage. Khaitan and Raju [2012] analyze the use of a sodium alanate storage system to store hydrogen for CAES. Yet, to store large amounts of gaseous fuels underground storage options are in general the least expensive technology coming along with large capacities in the TWh-scale. While natural gas can be stored in salt caverns and porous rock formations such as aquifers and depleted gas fields, hydrogen is rather constraint to salt caverns, since porous rock storage can lead to a high diffusivity losses and contamination (and reaction) of hydrogen with gases and minerals contained in porous rock [Crotogino and Hamelmann, 2007; Crotogino et al., 2010; Wolf, 2011]. The feasibility of hydrogen storage in salt caverns is proven in operation in Teesside (Sabic Europe, UK) and in Texas (ConocoPhillips and Praxair, US) [Crotogino and Hamelmann, 2007; Crotogino et al., 2010]. Detailed investigations to estimate the hydrogen underground storage potential are conducted, e.g. the European Project "HyUnder" [*HyUnder*] or the Germany project "InSpEE" [Horvath, Donadei, and Schneider, 2016; Donadei and Zander-Schiebenhöfer, 22./23. April 2015; Donadei et al., 2015; Pollok et al., 2015]. Horvath, Donadei, and Schneider [2016] estimate that the storage potential of hydrogen in underground salt caverns in Northern Germany is around 1614 TWh [Horvath, Donadei, and Schneider, 2016].

Böttcher et al. [2017] investigate the storage of hydrogen in salt caverns. They [Böttcher et al., 2017] use a thermodynamic approach to estimate temperatures and pressures during charging and discharging according to Xia et al. [2015] and combine these estimates with geo-mechanical stability analysis.

Khaledi et al. [2016] use Khaitan and Raju [2012] methods to describe the thermodynamics of a salt cavern air storage to investigate salt rock geo-mechanic stability.

Conclusion. The corollary of the above literature survey is that dynamic simulation of CAES with renewables mainly focus on economic aspects e.g. [Greenblatt et al., 2007; Mason and Archer, 2012; Mauch, Carvalho, and Apt, 2012; Gu et al., 2013; Madlener and Latz, 2013; Lund and Salgi, 2009; Drury, Denholm, and Sioshansi, 2011]. Time-dependent calculation of the thermodynamics of air storage caverns is largely explored without linking it to an overall process analysis [Budt et al., 2016a; Langham, 1965; Quast and Crotogino, 1979; Hoffeins, Romeyke, and Sütterlin, 1980; Quast, 1981; Crotogino, Mohmeyer, and Scharf,

2001-04-15; Raju and Khaitan, 2012; Kushnir, Ullmann, and Dayan, 2012a; Kushnir, Ullmann, and Dayan, 2012b; Kushnir, Dayan, and Ullmann, 2012; Khaitan and Raju, 2013; Xia et al., 2015; Zhao, Wang, and Dai, 2015; Marano, Rizzo, and Tiano, 2012; Maton, Zhao, and Brouwer, 2013; Hartmann et al., 2012a]. Different TES options are suggested and elaborated in detail including experimental validation [Zunft et al., 2012; Moser et al., 2012; Zunft, 2015; Bullough et al., 2004; Zunft et al., 2006; Jakiel, Zunft, and Nowi, 2007; Alstom Power et al., 2007; Dietz, 2008; Mei et al., 2015; Wolf, Berthold, and Dötsch, 16.06.2009; Wolf, 2011; Qi, 2012; Park et al., 2014]. Part load, start-up, and rundown procedures on the other hand have been addressed rarely [Nielsen, 2013; Mazloum, Sayah, and Nemer, 2016; Sciacovelli et al., 2017] and have not yet been compared to measured data. Finally, a global analysis containing all four of these transient aspects has been effected using several simplifications for an adiabatic CAES system only [Mazloum, Sayah, and Nemer, 2016; Sciacovelli et al., 2016] and, thus, will be discussed in this thesis in more detail for several CAES plant layouts including a discussion of hydrogen as fuel. Furthermore, a comprehensive set of measured operational data of the Huntorf CAES process is a unique feature to validate the here presented calculation methods.

1.3 Scope of this Thesis

The scope of this thesis is to describe and analyze existing and forthcoming CAES concepts considering air as a real gas and the processes as irreversible at steady state as well as time-dependent. Both energy concepts, enthalpy and exergy, are used to assess the characteristics of the different processes. Transient thermodynamic processes involved in CAES are analyzed in detail. The calculation methods are validated using a unique set of measured operational data from the Huntorf CAES plant. A new process using hydrogen to substitute natural gas and, thus, reduce green house gas emission is developed based on the steady state and dynamic thermodynamic models. This next generation CAES system is tested in a model 100 % renewable energy system environment. The uniqueness of the here presented calculation is the validation of the calculation methods with operational data from Huntorf CAES plant and a unified, thus, comparable analysis of different processes. In depth investigation on the technical efficiency of CAES are presented as it appeared that literature values are unambiguous. Furthermore, a novel economic equivalent-efficiency coefficient η_{eees} to rank fuel-driven CAES together with strictly electrical energy storage facilities is presented.

Chapter 2

Compressed Air Energy Storage – Basic Principles

[1] Fig. 2.1 shows a general concept of CAES with an associated temperature-entropy (T-s) diagram. The essential elements of CAES are: an electrical motor-generator (M/G), an air compressor (C), a compressed air storage (S), a burner (B), and a gas turbine (T). When a

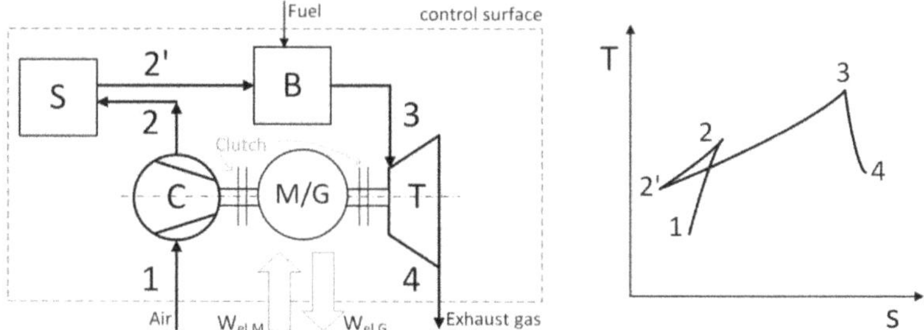

FIGURE 2.1: CAES open circuit and T-s diagram (with M/G- electrical motor/generator, C- compressor, S- compressed air storage, B- burner, T- turbine), adapted from [U.S. National Research Council, 1977; Giramonti and Lessard, 1974]

surplus of electricity occurs, the electrical motor (M) drives the air compressor (C) and the compressed air is then stored in the storage place (S). The electrical work needed to drive the compression is marked in Fig. 2.1 as $W_{el,M}$. On re-powering, a gaseous fuel is burned in the burner (B) and the high-pressure combustion products expand in the turbine (T) which drives the electricity generator (G). The work produced in the generator is marked in Fig. 2.1 as $W_{el,G}$. The two clutches allow for coupling the motor-generator with either the compressor or with the gas turbine. In the T-s diagram shown in Fig. 2.1, path 1-2 represents air compression, path 2-2′ indicates air storage, 2′-3 shows combustion whilst 3-4 represents expansion. From the thermodynamic point of view the CAES is a non-cyclic open-circuit process with air, fuel, and exhaust gas stream, as well as electrical work, crossing the control surface. More precisely, Fig. 2.1 shows two distinct processes: The first one, marked by path 1-2-2′, is a conversion of the electrical energy (work $W_{el,M}$) into compressed air energy and its storage (in what follows also referred to as "charge" mode) whilst the second process, marked by 2′-3-4 path, is a conversion of both the stored compressed air energy and the fuel chemical

[1]Parts of this Chapter have previously been published in [Kaiser, Weber, and Krüger, 2018].

energy into electrical energy (work $W_{el,G}$) ("discharge" mode). Such a distinction is useful since the processes 1-2-2′ and 2′-3-4 do not proceed simultaneously. The CAES shown in Fig.

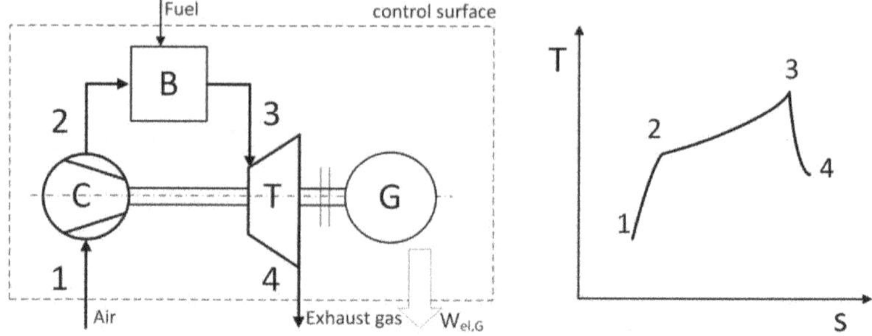

FIGURE 2.2: Open-circuit gas turbine and T-s diagram, adapted from [Borgnakke and Sonntag, 2009; Lechner and Seume, 2010]

2.1 is somewhat similar to an open-circuit gas turbine plant whose simplified circuit is shown in Fig. 2.2 together with an associated T-s diagram; path 1-2 shows the compression, path 2-3 combustion while path 3-4 indicates expansion. In a gas turbine open-circuit, air-stream and fuel-stream enter the control surface while combustion products and electrical work ($W_{el,G}$) leave the control surface. The essential difference to CAES is that the compression, combustion, and expansion proceed simultaneously. Thus, the electrical work leaving the control surface of an open-circuit gas turbine power plant is the difference between expansion and compression work. For example, for a typical gas turbine plant producing 100 MW power, the turbine would generate around 250 MW of which around 150 MW would be needed to run the compressor. Hence, decoupled operation of compressor and turbine is an important feature of CAES making a large power span available (in the example -150 to +250 MW power, which quadruples the power range of a conventional 100 MW gas turbine, from a grid point of view). The CAES process, such as the one shown in Fig. 2.1, is furthermore inherently time-dependent since during charging the pressure, temperature, and to a certain degree even the air flow rate, vary with time. During re-powering variations in the pressure, temperature, air flow rate, fuel gas flow rate do occur. Thus considerations and comparisons of different CAES concepts should include time-dependent simulations in which the thermodynamics of charging and discharging the storage should be properly handled.

2.1 Technical Work

In discussing performance criteria the concept of specific technical work is used whose inexact differential is defined as

$$\delta w_t = v \cdot dp \tag{2.1}$$

and

$$w_t = \int_{initial}^{final} v \cdot dp \tag{2.2}$$

In German literature on technical thermodynamics [Lechner and Seume, 2010; Baehr and Kabelac, 2009; Weber and Weber, 2010] the above expression represents so called specific "technical" work (in J/kg) as opposed to volume work, hence the subscript "t". Upon compression, the technical work is positive whilst it is negative upon expansion. The first law of thermodynamics is used here in the following formulation applicable to irreversible processes:

$$dh = \delta q + \delta w_t + \delta w_{friction} = Tds + \delta w_t \tag{2.3}$$

where dh and ds are the exact differentials of enthalpy and entropy, respectively; δq and $\delta w_{friction}$ are the inexact differentials of heat provided to the system and the inexact differential of friction work, respectively. If, in the thermodynamic analysis that follows, the processes are treated as reversible the friction work is omitted: $\delta w_{friction} = 0$. The thermodynamic analysis of the various CAES concepts in this thesis is carried out using the Engineering Equation Solver (EES). Gases are treated as real. For air, the non-dimensional Helmholtz equation of state [Lemmon et al., 2000] is used and enthalpy and entropy are calculated using differentiation with respect to density and temperature.

2.2 Exergy

[2] Sadi Carnot describes 1824 the upper limit to the work that can be done by a heat engine [Carnot, 1825]. This consideration can be seen as basis for the definition of exergy. In 1873, Josiah Willard Gibbs presents the mathematics of "available energy of the body and medium" which is today termed "exergy" [Gaggioli, Richardson, and Bowman, 2002] after a word creation by Zoran Rant in 1956.

Exergy is a means to quantify the amount of maximum useful work that can be carried out by a system in a specific environment. This concept can be applied to state variables and process variables. Some energy forms such as potential, kinetic, and electric energy as well as technical work and useful work [Baehr and Kabelac, 2009] can be completely transformed into useful work in reversible processes and can, thus, be entirely considered as exergy. Other energy forms can only be transformed partly into useful work such as heat, inner energy or enthalpy. Their transformability is limited by the state condition of the surroundings (subscript "∞"). Hence, exergy is a combined property of the system and its environment [Baehr and Kabelac, 2009]. Exergoeconomic analysis in general are investigated by Tsatsaronis [1993]. In this work the thermodynamic environment is considered as a stationary system which is in mechanical, thermal, and chemical equilibrium and whose intensive properties are not changed by the system.

In contrast to energy, which is a conservative property, exergy can be destroyed. Then, it turns into anergy which is defined as the part of energy that cannot serve to deliver useful work i.e. that is in complete equilibrium with the environment. The sum of exergy and anergy remains constant. Hence, the first law of thermodynamics for exergy can be formulated as: $exergy + anergy = const.\ (1^{st}\ law)$

The second law of thermodynamics applied to the concept of exergy says that only in reversible processes exergy remains constant. In irreversible processes exergy turns into anergy. These exergy losses (ex_{loss}) have to be taken into account in energy balances as: $exergy_{in} = exergy_{out} + ex_{loss}$ or in other words $ex_{loss} \geq 0\ (2^{nd}\ law)$

[2] Parts of this Chapter have previously been published in [Kaiser and Krüger, 2019].

Exergy of a stream of fluid that is in chemical equilibrium (otherwise chemical potential (μ) must be added) with the thermodynamic environment is: [Baehr and Kabelac, 2009] $ex = h - h_\infty - T_\infty(s - s_\infty) + 1/2(c^2 - c_\infty^2) + g(z - z_\infty)$

If kinetic and potential energy are negligible the exergy of a stream of fluid can be simplified to the specific exergy of enthalpy: [Baehr and Kabelac, 2009]

$$ex = h - h_\infty - T_\infty(s - s_\infty) \tag{2.4}$$

where subscript ∞ denotes ambient conditions.

Exergy (ex_q) of the process heat (q) [Doering, Schedwill, and Dehli, 2008] is given in Eq. (2.5) where T_i is the temperature of the heat flow crossing the control volume border [Baehr and Kabelac, 2009]:

$$ex_{qi,i-1} = (1 - T_\infty/T_i)q_{i,i-1} \tag{2.5}$$

Since exergy is not a conservative property, exergy balances of irreversible processes include a term for exergy losses. These exergy losses have to be calculated according to the type of system. In the following, two open system irreversible process examples, typical for CAES, are given.

Exergy balance of compressor or turbine. A fluid stream enters a compressor (or turbine) at state i-1 and leaves it at state i. Technical work ($w_t = \int vdp$) is added to (or removed from) the fluid stream. If the velocities and height of the air streams are equal ($c_{i-1} = c_i; z_{i-1} = z_i$), the technical work ($w_t$) can be determined as a difference of enthalpies ($h_{i-1} - h_i$) according to the energy balance of the system. The electrical work of the motor is slightly higher than the technical work due to mechanical losses estimated via the mechanical efficiency (η_{mech}). Thus, the energy balance can be formulated as: $0 = w_t + h_{i-1} - h_i = \eta_{mech} \cdot w_{el} + h_{i-1} - h_i$

To find the exergy loss ex_{loss} of such a process the exergy balance and the exergy of enthalpy (Eq.2.4) is used:

$ex_{loss} = w_t + ex_{hi-1} - ex_{hi}$

$ex_{loss} = h_i - h_{i-1} + h_{i-1} - h_\infty - T_\infty(s_{i-1} - s_\infty) - (h_i - h_\infty - T_\infty(s_i - s_\infty)$

$ex_{loss} = T_u(s_i - s_{i-1})$

Fig. 2.3 illustrates the system and its energy and exergy flows.

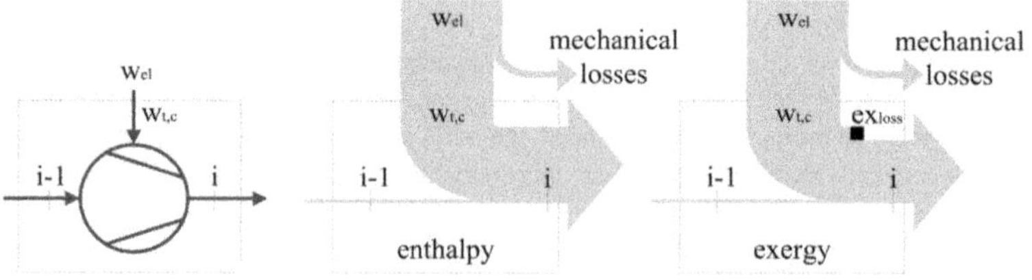

FIGURE 2.3: Energy and exergy balance of a compressor unit

Thus, the exergetic losses of the compressor amount to [Baehr and Kabelac, 2009]:

$$ex_{loss,wt,i,i-1} = T_\infty(s_{i-1} - s_i) \tag{2.6}$$

A similar derivation applies to turbine stages.

Exergy balance of cooling and combustion stages. In cooling (or heating) stages a heat flow rate q is removed from (or added to) the fluid stream. The energy balance shows that the heat removed (or added) corresponds to the enthalpy difference of the fluid. Energy balance: $0 = h_{i-1} - h_i - q$

For exergy on the other hand, the useful work, and thus exergy, contained in the heat (q) is limited by the Carnot-factor. Thus, the exergy of the air stream at $i-1$ is only stripped by an exergetic heat of $ex_q = q(1 - T_\infty/T_i)$. Again, the exergy of the air streams is calculated with Eq. 2.4 and an exergy loss for heat transfer $ex_{loss,q}$ has to be taken into account. Thus, the exergy balance is: $ex_{loss,q} = ex_{hi-1} - ex_{hi} - ex_q$

$ex_{loss,q} = h_{i-1} - h_\infty - T_\infty(s_{i-1} - s_\infty) - (h_i - h_\infty) - T_\infty(s_i - s_\infty) - (h_{i-1} - h_i)(1 - T_\infty/T_i)$

$ex_{loss,q} = T_\infty(s_i - s_{i-1}) + T_\infty/T_i(h_{i-1} - h_i)$

Thus, the exergy losses of a heat exchanger or, with a similar derivation, combustion stage amount to [Borgnakke and Sonntag, 2009]:

$$ex_{loss,q,i,i-1} = T_\infty(s_i - s_{i-1}) + T_\infty/T_i(h_{i-1} - h_i) \tag{2.7}$$

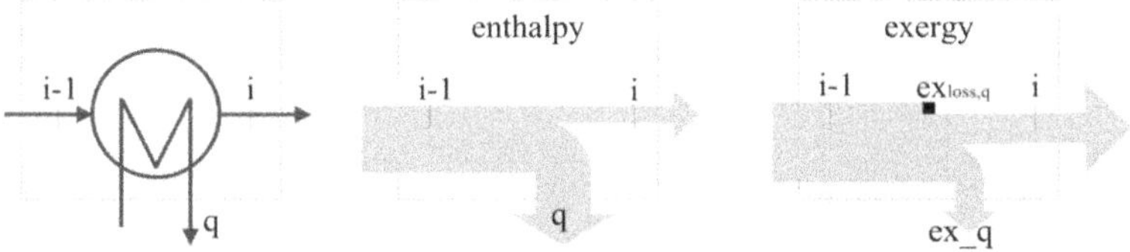

FIGURE 2.4: Energy and exergy balance of a heat transfer unit

2.3 Efficiency

Generally, efficiency of a process or a machine is defined as the ratio of useful output (product) to the efforts put into producing the output. For thermodynamic processes it is then the ratio of useful work produced to the energy input:

$$\eta = \frac{useful\ work\ produced}{energy\ input}$$

Application of this definition to heat engines is rather straight forward since both the useful work produced and the energy input are easy to define (Fig. 2.5(a)). Similarly, when electrical energy is stored in batteries, in a Pumped Hydro Energy Storage system or in an Adiabatic CAES system (Fig. 2.5(b)), there are no ambiguities in defining the terms appearing in the above definition. Problems arise when the above definition is applied to a CAES system where the goal is to store electrical energy and, in order to carry out such a storage, a fuel input is needed without which the storage is not realizable at all (Fig. 2.5(c)). Then, the question arises how to handle this extra fuel input when efficiency is to be calculated.

CAES efficiency. In the context of CAES, an efficiency (η_{caes}) defined as

$$\eta_{caes} = \frac{|W_{el,G}|}{W_{el,M} + Q_{fuel}} \tag{2.8}$$

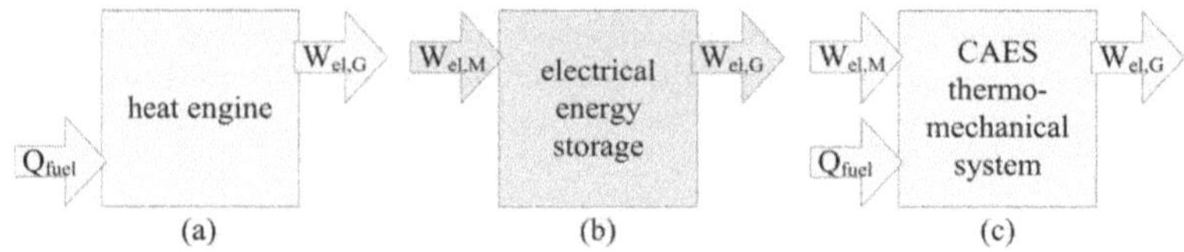

FIGURE 2.5: Input and output streams required for efficiency definition of different types of processes (a) heat engines, (b) electrical energy storage systems, and (c) thermo-mechanical systems such as CAES

is commonly used with the electrical work generated ($W_{el,G}$) and consumed ($W_{el,M}$), respectively, and the fuel enthalpy (Q_{fuel}), as depicted in Fig. 2.1 or 2.5(c).

Even though η_{caes} as defined in Eq.(2.8) is very simple and stringent, it is limited to a comparison of thermo-mechanical processes that are having both, electrical and fuel energy inputs. It is not suitable for a comparison with electrical energy storage efficiencies e.g. of PHES, batteries, or adiabatic (fuel-free) CAES.

Thermal efficiency. By analogy to fuel-driven energy conversion processes (heat engines), a thermal efficiency (η_{th}) can be defined as [Borgnakke and Sonntag, 2009]:

$$\eta_{th} = \frac{|W_{el,G}| - W_{el,M}}{Q_{fuel}} \tag{2.9}$$

Such defined thermal efficiency (η_{th}) allows a comparison of CAES plants to other heat engines, like conventional gas turbines. However, one has to point out that the electrical energy used for the compression may originate from renewable energy sources and then the net value of turbine work minus compression work used as numerator in Eq.(2.9) is somewhat misleading. Obviously, Eq.(2.9) in only applicable when $Q_{fuel} > 0$ and is therefore not applicable to (fuel-free) electrical energy storage systems.

Heat rate. The heat rate defined as

$$hr \begin{cases} hr_1 = \frac{1}{\eta_{th}} = \frac{Q_{fuel}}{W_{el,G}-W_{el,M}} \\ hr_2 = \frac{Q_{fuel}}{W_{el,G}} \; when \; W_{el,M} \; is \; omitted \end{cases} \tag{2.10}$$

can be applied to heat engines, CAES and strictly electrical energy storage technologies with little constraints. The heat rate is the amount of fuel energy used per electrical energy supplied to the grid and often expressed in $[\frac{kWh_{fuel}}{kWh_{electric}}]$. Conventional gas turbines have characteristic values of $hr_1 = \frac{Q_{fuel}}{W_{el,G}-W_{el,M}} = 2.7\frac{kWh_{fuel}}{kWh_{electric}}$ (e.g. Siemens SGT-800). For CAES the compression work, that is ideally driven by a surplus of renewable power is omitted. Hence, the heat rate of the Huntorf CAES plant is $hr_2 = \frac{Q_{fuel}}{W_{el,G}} = 1.7\frac{kWh_{fuel}}{kWh_{electric}}$ (see Chapter 3), which is considerably lower than values applicable to gas turbines (this fact was highlighted in the early CAES-studies as a main asset of CAES [Bush et al., 1976]). McIntosh CAES reaches values as low as $hr_2 = 1.2$ (see Chapter 3). Obviously, for an electrical energy storage facility operating without fuel $hr = 0$ (see Table 2.1).

Round trip efficiencies. When considering CAES as a means for storing electrical energy, the round trip efficiency (η_{rt}) of electrical energy storage facilities, defined as a ratio of

Technology	heat rate in $[\frac{kWh_{fuel}}{kWh_{electric}}]$
Gas turbine Siemens SGT-800	2.7
CAES Huntorf	1.7
CAES McIntosh	1.2
ACAES	0
Pumped Hydro Energy Storage	0
Battery	0

TABLE 2.1: Heat rate (hr) for several energy storage technologies in comparison with Siemens SGT-800 turbine

electrical energy supplied to the grid during discharging to energy taken from the grid during charging [Kaiser and Busch, 2015; Succar and Williams, 2008], may be introduced:

$$\eta_{rt} = \frac{electrical\ energy\ supplied\ to\ the\ grid}{electrical\ energy\ taken\ from\ the\ grid}$$

However, this semantically simple approach does not have much physical sense, since the ratio of $W_{el,G}$ to $W_{el,M}$, that is used to calculate η_{rt} of pumped hydro energy storage or batteries would, when applied to CAES, ignore the contribution of the fuel energy (compare with Fig. 2.1 or 2.5(c)) and, thus, lead to values > 1.

$$\eta_{rt1} = \frac{|W_{el,G}|}{W_{el,M}}(> 1 for\ fuel\ driven\ CAES) \tag{2.11}$$

For Huntorf CAES a value of $\eta_{rt1} = 119\,\%$ results. Thus, another calculation method has to be developed that distinguishes the contributions of the fuel and the electrical energies. This can be achieved in two ways: (**a**) by converting the fuel enthalpy into an electrical energy **equivalent** using a reference efficiency (η_{ref}) or (**b**) by calculating the **fraction** of the energy taken from the grid during charging that is returned to the grid during discharging.

For the first option (**a**), the following two definitions have been used [Kim et al., 2012; Nielsen, 2013; Succar and Williams, 2008; Elmegaard and Brix, 2011; Garvey, 2015]:

$$\eta_{rt2} = \frac{|W_{el,G}|}{W_{el,M} + Q_{fuel} \cdot \eta_{ref}} \tag{2.12}$$

and [Budt et al., 2016a; Succar and Williams, 2008; Steinmann, 2016]:

$$\eta_{rt3} = \frac{|W_{el,G}| - Q_{fuel} \cdot \eta_{ref}}{W_{el,M}} \tag{2.13}$$

In such calculations, the reference efficiency (η_{ref}) is a decisive factor [Budt et al., 2016a]. A wide range of possible values has been used in literature: $\eta_{ref} = 0.4$ [Kim et al., 2012; Steinmann, 2016; Elmegaard and Brix, 2011], 0.476 [Succar and Williams, 2008], 0.5 [Garvey, 2015] or 0.6 [Steinmann, 2016].

In option (**b**), the electrical energy taken from the grid ($W_{el,M}$) is reduced by the mechanical efficiency of the compressor and the heat losses during inter- and after-cooling, which gives the actual energy content of the stored air as $E_{air} = W_{el,M} \cdot \eta_{mech} - Q_{loss}$ (e.g. illustrated by the Sankey diagram of Huntorf's energy flows Fig.3.7). During discharging this energy is further reduced by the conversion efficiency of the turbine ($\eta_{tc} = \frac{W_{el,G}}{E_{air}+Q_{fuel}}$). The resulting value is set into relation with the amount of energy that was originally taken from the grid

($W_{el,M}$), hence $\eta_{rt4} = \frac{E_{air} \cdot \eta_{tc}}{W_{el,M}}$ which is equal to Eq.(2.14).

$$\eta_{rt4} = \frac{(W_{el,M} \cdot \eta_{mech} - Q_{loss}) \cdot (\frac{W_{el,G}}{W_{el,M} \cdot \eta_{mech} - Q_{loss} + Q_{fuel}})}{W_{el,M}} \tag{2.14}$$

In other words, η_{rt4} corresponds to the fraction of the electrical energy taken from the grid that is returned to the grid.

The so defined electrical energy storage round trip efficiency (η_{rt4}, Eq.(2.14)) results for Huntorf and McIntosh CAES in very low efficiency values of around 3 % (see Chapter 3). However, this calculation method (first presented in [Kaiser, Weber, and Krüger, 2018]) represents the actual flow of electrical energy through the CAES system in its physical sense. Eq.(2.14) can also be written as $\eta_{rt4} = \eta_{cc} \cdot \eta_{tc}$, where η_{cc} stands for the compressor conversion factor (in equivalence to η_{tc} for the turbines) and amounts to $\eta_{cc} = \frac{E_{air}}{W_{el,M}} = 0.05$ in the Huntorf and McIntosh examples. It is then apparent that as long as the heat removed during compression is wasted and not being recovered in the expansion, the so defined electrical energy storage efficiency will remain very low, indeed.

Round trip efficiency as an economic parameter. [3]Energy storage efficiency is an important tool to rank energy storage technologies in terms of profitability. The operational cost is mainly determined by energy cost, whereas the earnings depend on energy selling price, hence, the decisive factors determining economic viability are price differences between peak and off-peak electricity tariffs ('price spread') and the amount of energy that can be bought and sold. In other words the profits are determined by solving $profit(energy\ storage) = \Sigma(W_{el,G} \cdot p_{sell}) - \Sigma(W_{el,M} \cdot p_{buy})$, with p as electricity price (in [\$/kWh]). The upper and lower electricity price, p_{sell} and p_{buy}, respectively, vary depending on the market situation. To enable a general solution of the problem, it is assumed that the upper and lower prices can be represented by a mean buying and selling value, respectively.

The energy storage efficiency (η_{rt1} for strictly electric energy storage technologies) links both energy values, the amount of bought energy ($W_{el,M}$) and sold energy ($W_{el,G}$), Eq.2.11. The formulas can, thus, be transformed to the non-dimensional form $\frac{profit}{W_{el,G} \cdot p_{buy}} = \frac{p_{sell}}{p_{buy}} - \frac{1}{\eta}$. In other words, profits per energy units are positive as soon as the price spread factor ($\frac{p_{sell}}{p_{buy}}$) is larger than the reciprocal of the energy storage efficiency ($\frac{1}{\eta}$). Thus, the role of the energy storage efficiency (η) as important characteristic value to rank different storage technologies is obvious. However, due to the additional fuel cost this definition is not applicable to CAES. Hence, a new calculation method is developed:

Economic equivalent energy storage coefficient for CAES. The profits of a fuel-driven CAES plant not only depend on the electric energy prices but also on the amount of fuel (Q_{fuel}) and its price (p_{fuel} in [\$/kWh]) [Kaldellis and Zafirakis, 2007]. Hence, the profits can be determined by solving $profit(CAES) = \Sigma(W_{el,G} \cdot p_{sell}) - \Sigma(W_{el,M} \cdot p_{buy}) - \Sigma(Q_{fuel} \cdot p_{fuel})$. To eliminate $W_{el,M}$ and Q_{fuel} the well known process efficiency (η_{caes}, defined in Eq.2.8) and the heat rate (hr_2, Eq.2.10) can be used. Thus, the energy specific profits of CAES can be estimated as: $\frac{profit(CAES)}{W_{el,G} \cdot p_{buy}} = \frac{p_{sell}}{p_{buy}} - \frac{1}{\eta_{caes}} + hr_2 \cdot (1 - \frac{p_{fuel}}{p_{buy}})$. The **economically equivalent**

[3]The economic equivalent energy storage coefficient for CAES was first presented in a working paper [Kaiser, 2016]

energy storage efficiency for CAES (η_{eees}) is then defined as:

$$\eta_{eees} = \frac{1}{\frac{1}{\eta_{caes}} - hr_2 \cdot (1 - \frac{p_{fuel}}{p_{buy}})} \tag{2.15}$$

With a so defined efficiency, the non-dimensional earnings of CAES can be estimated via: $\frac{profit(CAES)}{W_{el,G} \cdot p_{buy}} = \frac{p_{sell}}{p_{buy}} - \frac{1}{\eta_{eees}}$ which corresponds to the calculation method applicable to strictly electrical energy storage units. Thus, the unambiguous electrical energy storage efficiency of strictly electric energy storage facilities can be replace in simplified economic comparisons to CAES plant by the here defined η_{eees} value.

For the two existing CAES plants Huntorf and McIntosh examples of such calculations of η_{eees} are presented in Table 2.2. Two price scenarios for a low gas price that is one third of the lower electricity price ($p_{fuel}/p_{buy} = 1:3$) and a second gas price that is as high as the electricity price ($p_{fuel}/p_{buy} = 1:1$) are given.

	η_{caes}	hr_{caes}	$\eta_{eees}(\frac{p_{gas}}{p_{buy}} = 1:1)$	$\eta_{eees}(\frac{p_{gas}}{p_{buy}} = 1:3)$
Huntorf	0.42	1.7	42 %	80 %
McIntosh	0.54	1.2	54 %	95 %

TABLE 2.2: Economic equivalent energy storage efficiency (η_{eees}) for the CAES plants Huntorf and McIntosh

The example in Table 2.2 shows that the Huntorf CAES plant, despite its seemingly low process efficiency ($\eta_{caes} = 0.42$), can - as long as gas prices are low - compete with modern PHES schemes that have efficiencies around 80 %. The McIntosh CAES plant having an elevated process efficiency of $\eta_{caes} = 0.54$ is with an economic equivalent energy storage efficiency of η_{eees} = 95 % at low gas prices more economic than PHES.

Thus, if the ratio of fuel price to electricity price is given, the economic equivalent energy storage efficiency for CAES (η_{eees}) can serve to easily rank different energy storage facilities in terms of economic viability. In Fig. 2.6 such a ranking is illustrated. CAES (at two different gas prices) is compared with two strictly electrical storage technologies: PHES and Adiabatic CAES (ACAES). In Fig. 2.6 the characteristic values of McIntosh ($\eta_{caes} = 0.54$ and $hr_{caes} = 1.2$) are used to represent the properties of CAES.

It has to be noted that the economic equivalent energy storage efficiency (η_{eees}) for CAES has *no physical* significance as it depends on economic parameters.

Exergy in efficiency considerations. In addition to the above discussed efficiencies based on enthalpy considerations, one can also use exergy to define these values. When using exergy, electric energy values such as $W_{el,G}$ and $W_{el,M}$ remain unchanged since electric energy is entirely convertible into useful work i.e. exergy. Differences between exergy efficiencies and enthalpy based considerations occur when heat Q is considered. The Carnot factor is then used to reduce the amount of heat considered. For these exergy efficiencies the symbol ξ is used.

Comparison and meaning of different efficiency values. In summary, it can be stated that there is no universal electrical energy storage efficiency for CAES. Every proposed efficiency definition has certain limitations. The drawbacks of these definitions are summarized in Table 2.3. Hence, the most revealing definition has to be chosen for specific applications; when

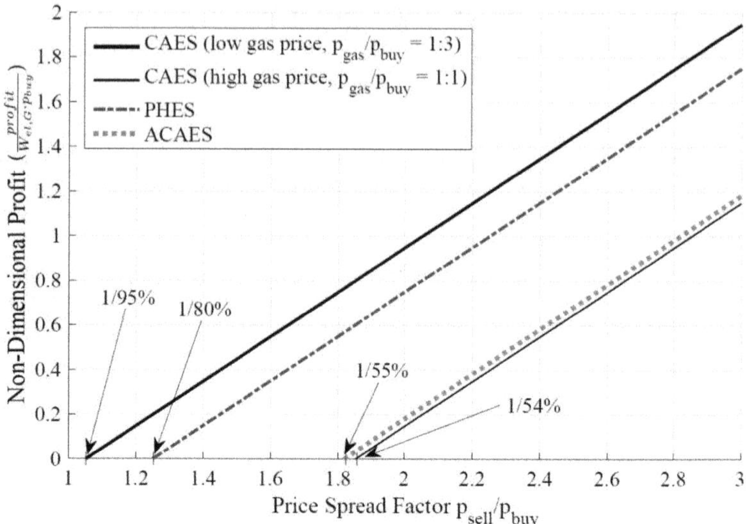

FIGURE 2.6: Non-dimensional profits of CAES, PHES, and ACAES

several thermo-mechanical concepts are compared η_{caes} can be used; when it comes to heat engines η_{th} and hr are useful.

TABLE 2.3: Drawbacks of the different efficiency definitions

Symbol	Eq.	Drawback
η_{caes}	2.8	limited to thermo-mechanical concepts
η_{th}	2.9	limited to heat engines
hr	2.10	applicable when fuel energy is used
η_{rt1}	2.11	limited to strictly electrical energy storage
η_{rt2}	2.12	includes the arbitrary factor η_{ref}
η_{rt3}	2.13	includes the arbitrary factor η_{ref}
η_{rt4}	2.14	enthalpy-based
η_{eees}	2.15	techno-economic value, function of gas price

Today's main focus is on energy storage applications, and thus, when it comes to comparing CAES to electrical energy storage devices, such as batteries or pumped hydro energy storage plants, the different meanings of the seemingly trivial electric energy storage efficiency have to be distinguished in order to assign the appropriate CAES counterpart. Table 2.4 lists several meanings of the electric energy storage efficiency. The example of Pumped Hydro Energy Storage (PHES) as one standard electric energy storage system in comparison with the corresponding CAES performance value is given. The electric energy storage efficiency of PHES amounts for instance to 80 %.

It appears that the electric energy storage efficiency can be interpreted in several ways. For PHES (and other electric energy storage units) these different meanings collapse to the use of a unique efficiency value (here $\eta(PHES) = 80$ %). However, to transfer these meanings to CAES several different characteristic values have to be used depending on the intended

TABLE 2.4: Electric energy storage efficiency and its different meanings for pumped hydro energy storage (PHES) with an actual energy storage efficiency η_{es} and Compressed Air Energy Storage (CAES) without unambiguous storage efficiency

Interpretation of electric energy storage efficiency	PHES	CAES
How much of the electrical input energy is recovered after the overall process?	η_{es}	η_{rt4} or ξ_{rt4}
How much electricity output can be generated per electrical energy input?	η_{es}	η_{rt1}
How much electricity output can be generated per overall energy input (sum of electrical and fuel)?	η_{es}	η_{caes} or ξ_{caes}
How much electrical energy is needed per output?	$1/\eta_{es}$	$1/\eta_{rt1}$
How much energy (sum of electrical, chemical and other energy forms) is needed to generate 1 kWh?	$1/\eta_{es}$	$1/\eta_{caes}$
How much of the electrical energy input is lost?	1-η_{es}	1-η_{rt4} or 1-ξ_{rt4}
How much of the overall energy input (sum of electrical, chemical and other) is lost?	1-η_{es}	1-η_{caes}
What is the minimum price spread factor required for a profitable operation if energy costs are the only costs considered?	$1/\eta_{es}$	η_{eees}

evidence. These are η_{rt4}, ξ_{rt4}, η_{rt1} $(= \xi_{rt1})$, η_{caes}, and ξ_{caes} as listed in Table 2.4. Furthermore, if economic considerations are concerned, an additional value termed η_{eees} is introduced. This economic equivalent energy storage coefficient for CAES is a characteristic value of a CAES process and the energy cost of fuel and electric energy. Hence, it is no technical value but a techno-economic value which enables to rank CAES with other electric energy storage systems based on a simplified profitability check [Kaiser, 2016].

Conclusions. While the efficiency of adiabatic CAES is unambiguously defined as η_{rt1}, a multitude of possible efficiency values can be defined for fuel-driven CAES. Despite the fact that $\eta_{caes} = W_{el,G}/(W_{el,M} + Q)$ is a widespread method to characterize fuel-driven CAES, it is unsuitable to characterize CAES as energy storage unit because the denominator mixes two different types of energy input (fuel and electricity). Several approaches to rectify this mismatch via reference efficiency can be regarded as imprecise [Elmegaard and Brix, 2011; Budt et al., 2016a]. Hence, its exergetic counterpart ξ_{caes} is also redundant. $\eta_{rt1}(= \xi_{rt1})$ results for fuel-driven CAES in values larger than 1 which disagrees with the basic formulation of efficiency. Nevertheless, this coefficient can be useful in the above mentioned narrower sense to quantify 'how much electricity output can be generated per electric energy input'. Yet, such a coefficient should be supplemented by fuel demand of the process, e.g. in the form of a heat rate value. Thus, the primary meaning of electric energy storage efficiency, which is "how much of the electric energy input can be recuperated as electric energy output", is only represented by both values η_{rt4} and ξ_{rt4}. Considering the general definition of efficiency, the exergy-based ξ_{rt4} value, or more precisely $\xi_{cc} \cdot \xi_{tc} = \xi_{rt4}$, is preferred as it allows for a more intuitive use of efficiency for the energy content of compressed air. However, when CAES is not only considered as a storage unit but also a generation unit, enthalpy-based approaches seem handier because they allow for easier estimates of fuel consumption. Thus, the parameters $\eta_{cc} \cdot \eta_{tc} = \eta_{rt4}$ are deemed most helpful. Further, η_{tc} reflects the superior generation characteristics of CAES as one main asset of this fuel conversion technology.

The discharging of CAES can be considered as a fuel-saving generation of electrical energy with efficiencies that are higher than state of the art combined gas and steam turbines. In any way, calculation assumptions such as the mechanical efficiency η_{mech} and the inner efficiency η_s must be displayed to make calculation methods retrievable. For techno-economic considerations a more complex approach, such as η_{eees}, is inevitable.

Chapter 3

Steady State Thermodynamics of CAES

[1] Currently, there are two commercial CAES plants in operation: the Huntorf plant near Bremen (North Germany) and the McIntosh plant in Alabama, U.S. Some basic data is given in Table 3.1. Huntorf, in operation since 1978, was originally operated by Nordwestdeutsche

Location	Commissioning	Compressor	Turbine	Cavern Volume
Huntorf	1978	68 MW	321 MW	$310,000\,m^3$
McIntosh	1991	49 MW	110 MW	$538,000\,m^3$

TABLE 3.1: Basic data of CAES plants Huntorf and McIntosh [Crotogino, Mohmeyer, and Scharf, 2001-04-15; Pollak, 1994; Krüger, 27.07.2015]

Kraftwerke AG and belongs nowadays to Power Plants Group Wilhelmshaven of Uniper Kraftwerke GmbH (formerly E.ON Kraftwerke GmbH). The site was designed as a back-up and emergency plant for the electricity supply system (grid) of North Germany. Its location was due to availability of salt caverns in this region. In the case of a black-out, the plant should use the stored compressed air to initiate the electricity generation process to bring the collapsed system back to operation (black start ability). Such an emergency has not occurred yet. McIntosh, commissioned in 1991, was originally owned by Alabama Electric Cooperative, Inc. and constructed by Harbert International and Gibbs and Hill (joint venture) [Pollak, 1994]. Nowadays, it is operated by PowerSouth Energy Cooperative and used as a power plant for peak demand and storage facility benefiting from electricity price differential between day and night.

3.1 Huntorf Plant

Fig. 3.1 shows a process flow diagram of the Huntorf plant with a subdivision into two operation modes 'Charge' and 'Discharge'.

Charge. For charging, a compression with several stages is used since the pressure ratio ($\frac{p_8}{p_1}$) is high varying in the overall range of 20:1 to 68:1 corresponding to the minimum ($20\,bar$) and maximum ($68\,bar$) cavern pressures. The compression is divided into a low pressure and a high pressure unit. The low pressure compressor (C_I) is a one stage axial compressor followed by a cooling stage. The high pressure compressor is a six stage centrifugal unit with

[1] Parts of this Chapter have previously been published in [Kaiser, Weber, and Krüger, 2018].

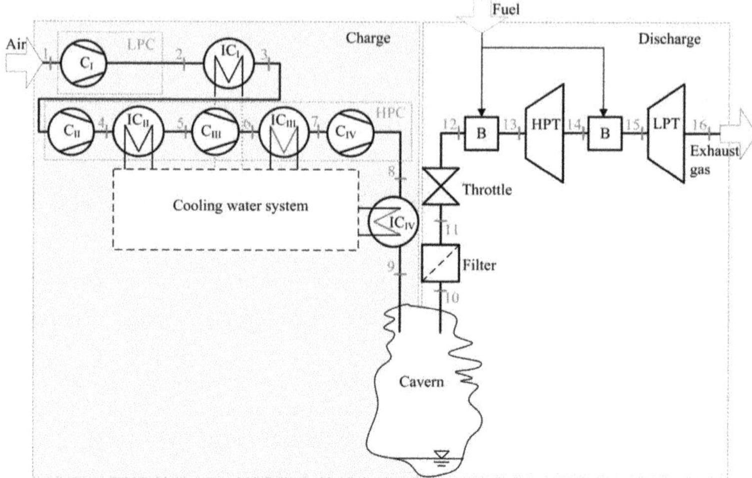

FIGURE 3.1: Process flow diagram of Huntorf CAES, subdivided into 'Charge' and 'Discharge' (C - compressor; T - turbine; B - burner), adapted from [Brown Boveri & Cie, 1980; Crotogino, Mohmeyer, and Scharf, 2001-04-15]

cooling after every two stages [Hoffeins, Romeyke, and Sütterlin, 1980]. Thus, each two high pressure compressor stages are treated as one unit, as indicated in Fig. 3.1 (C_{II}, C_{III}, C_{IV}). Inter-cooling between the compressor stages reduces the work required for compression (IC_I to IC_{III} in Fig. 3.1). After the compression, a cooling (IC_{IV}) is required to cool down the air to the permissible inlet temperature of the cavern ($50°C$). During charging the air mass flow rate is kept at $108\ kg/s$ [Hoffeins, Romeyke, and Sütterlin, 1980; Crotogino, Mohmeyer, and Scharf, 2001-04-15].

Cavern. The compressed air is stored in two solution-mined salt caverns of around $140,000$ and $170,000\ m^3$ volumes [Crotogino, Mohmeyer, and Scharf, 2001-04-15]. Typically both caverns are used simultaneously as if it were a single storage volume [Krüger, 27.07.2015]. The advantage of two separate caverns appears when one cavern is emptied to atmospheric pressure for repair or maintenance. The second cavern is then used to refill the first one to a minimum pressure ($> 20\ bar$), that is required to operate the compressor train [Crotogino, Mohmeyer, and Scharf, 2001-04-15; Krüger, 27.07.2015]. Without this option an additional compressor would have to be used to refill the empty cavern.

Discharge. In discharging mode, the air mass flow rate amounts to $455\ kg/s$ in a full load operation. The air is filtered, then passes through a throttle, and is expanded in two turbines: a high pressure turbine (HPT) and a low pressure turbine (LPT), each with supplementary firing of natural gas. If the cavern pressure is in the 46 to 68 bar range, the turbines are operated at full load.

Thermodynamic data. In Fig. 3.1 and Table 3.2 numbers $i = 1$ to 16 indicate the thermodynamic states of the process. The design parameters [Brown Boveri & Cie, 1980; Hoffeins, Romeyke, and Sütterlin, 1980; Crotogino, Mohmeyer, and Scharf, 2001-04-15; Krüger, 27.07.2015] are listed in Table 3.2 using bold face font. Table 3.2 contains also several sets of measured data (temperature, pressure, flow rates): four sets for charging (indicated in Table 3.2 by t=1,2,3,4, see also Table A.1) and three sets for discharging (t=5,6,7) [Krüger, 27.07.2015].

i		Pressure (p) in bar					Temperature (T) in K			
Charge		t=1	2	3	4		t=1	2	3	4
1		0.99	0.99	0.99	0.99		283	283	283	284
2	**6**	6.06	6.14	6.16	6.27		508	511	512	514
3		5.90	5.96	6.00	6.10	**308**	304	305	305	306
4		13.94	14.19	14.28	14.71		421	423	423	426
5		n.a.	n.a.	n.a.	n.a.	**308**	308	309	308	310
6		27.95	28.82	29.34	30.70		400	401	402	404
7		n.a.	n.a.	n.a.	n.a.	**309**	308	309	309	310
8	***68**	57.15	59.90	62.29	66.70		417	420	421	425
9		55.77	59.09	60.93	65.34	**322**	322	324	323	325
Discharge		t=5	6	7			t=5	6	7	
10		50.4	52.1	53.6			n.a.	n.a.	n.a.	
11		47.42	n.a.	n.a.			307	309	309	
12	**42**	n.a.	n.a.	n.a.			304	303	301	
13	**41.3**	38.48	28.62	18.83		**763**	774	786	805	
14		11.8	8.13	46.00			583	583	583	
15	**12.8**	n.a.	n.a.	n.a.		**1218**	1217	1095	978	
16		1.01	1.01	1.01			747	704	663	
Flow Rate and LCV										
	air in charge mode (t=1 to 4)						**108 kg/s** (108 kg/s)			
	air in discharge mode (t=5/6/7)						**455 kg/s** (402/301/200 kg/s)			
	fuel (natural gas)						13.2 kg/s			
	LCV of natural gas						41 MJ/kg			

TABLE 3.2: Operation parameter of Huntorf CAES plant with **design data** [Brown Boveri & Cie, 1980] in boldface font and several sets of measured operation data [Krüger, 27.07.2015] ordered by state point number i in conjunction with Fig. 3.1. (*non-steady-state value with minimal permissible limit of 20 bar at plant operation in a part load; LCV = lower calorific value; n.a. = not available)

3.1.1 Inner Efficiency

Charge. The measured data are used to determine the inner (thermodynamic) efficiency of compression (often referred to as isentropic efficiency). Since temperatures and pressures have been measured at points 1 and 2 (see Table 3.2), the inner efficiency (η_s) is obtained upon solving the equations

$$s(T_1, p_1) = s(T_{2s}, p_2) \tag{3.1}$$

and

$$\eta_s(LPT) = \frac{h(T_{2s}, p_2) - h(T_1, p_1)}{h(T_2, p_2) - h(T_1, p_1)} \tag{3.2}$$

where s(T,p) and h(T,p) are appropriate functions for specific entropy and specific enthalpy of air treated as a real gas; T_{2s} represents the temperature of isentropic compression. The arithmetic average of the four sets of measured data ($t = 1, 2, 3, 4$) gives $\eta_s(LPC) = 0.844$ which corresponds to $T_2 = 502\ K$; typical values for η_s lie within 0.70 to 0.88 range [Borgnakke and Sonntag, 2009] or for the newest turbo compressors even within 0.86 to 0.90 range [Lechner and Seume, 2010; Baehr and Kabelac, 2009]. The inner efficiencies η_s of the HPC stages are obtained in the same way and the results are given in Table 3.3. To reduce the technical work needed for the overall compression, each compression stage is followed by an inter-cooler (IC_I to IC_{III} in Fig. 3.1) [Brown Boveri & Cie, 1980; Krüger, 27.07.2015]. The compressed air is cooled down to $T_3 = 308\ K$ using water and the pressure stays nearly constant at $6\ bar$. Yet, a small pressure loss occurs and is estimated, using procedures applicable to tubular heat exchangers [VDI, 2013], to be $10\ mbar$. The inter-staged pressures of C_{II}, C_{III} and C_{IV} vary with the cavern pressure since the overall compression ratio ($\frac{p_8}{p_2}$) varies from 20:6 (empty caverns) to 68:6 (full caverns). It is assumed, that the inter-stage pressures of the HPC correspond to those pressures that lead to a minimum overall technical work. Thus, for a three-staged compression, from $6\ bar$ to the maximum cavern pressure of $p_8 = 68\ bar$, the compression ratio β is:

$$\beta = (\frac{p_8}{p_3})^{1/3} = (\frac{68}{6})^{1/3} = 2.246 \tag{3.3}$$

and the inter-staged pressures are then $p_4 = 13.5\ bar$ and $p_6 = 30.3\ bar$. The temperature and pressure values at points 4 to 8 are calculated following the same procedure (see Eq.(4), (5)). The pressure losses in the inter-coolers increase with the pressure level, see Table 3.4. The after-cooler (IC_{IV} in Fig. 3.1) makes sure that the temperature of the compressed air does not exceed the maximum allowable temperature of the cavern (323 K), thus, the compressed air is cooled to $T_9 = 322\ K$. A pressure loss of $800\ mbar$ is estimated.

	Process Unit	Inner Efficiency η_s
Charge	C_I	0.844
	C_{II}	0.726
	C_{III}	0.764
	C_{IV}	0.653
Discharge	HPT	0.894
	LPT	0.894

TABLE 3.3: Inner efficiencies (η_s) of the compressors (C) and turbines (T) calculated using measured Huntorf operation data [Krüger, 27.07.2015]

Discharge. After leaving the cavern the air is filtered. Due to a high air velocity (up to 30 m/s [Quast, 1981]) at the cavern exit and due to the filters, a pressure loss of around $4\ bar$ occurs during full load operation so the maximum pressure at the filter outlet is $p_{11} = 64\ bar$ [Krüger, 27.07.2015]. Further reduction to $p_{12} = 42\ bar$ pressure is then caused by a throttle so as to keep, under consideration of a pressure loss inside the combustion chambers of approximately $0.7\ bar$, a constant pressure of $p_{13} = 41.3\ bar$ at the turbine inlet at full load operation. The throttling is an isenthalpic pressure drop that comes along with a temperature change (Joule-Thomson-Effect). To calculate the outlet temperature T_{12} of the throttle, the equation $h(T_{11}, p_{11}) = h(T_{12}, p_{12})$ is solved with an appropriate function for the enthalpy of air treated as a real gas. If the cavern pressure drops below $46\ bar$, no more throttling is used and the turbine starts to operate in a part load.

In the burner, natural gas is injected into the throttled air to increase the temperature to the design value of $T_{13} = 763\ K$ so as to avoid icing. The outlet pressure of the HPT (p_{14}) corresponds to the inlet pressure of the LPT (p_{15}) plus the pressure loss inside the second combustion chamber that is set equal to the pressure loss of the first combustion chamber $(0.7\ bar)$. Since the outlet temperature of the turbine (T_{14}) has been measured (see Table 3.2) the inner efficiency of the HP expansion can be calculated as

$$\eta_s(HPT) = \frac{h(T_{13}, p_{13}) - h(T_{14}, p_{14})}{h(T_{13}, p_{13}) - h(T_{14s}, p_{14})} \tag{3.4}$$

where T_{14s} is the outlet temperature of the turbine under reversibility assumption which is obtained upon solving $s_{13}(p_{13}, T_{13}) = s_{14}(p_{14}, T_{14s})$ equation. The 0.894 value for $\eta_s(HPT)$ listed in Table 3.3 is an arithmetic average of the HPT expansion efficiencies derived using the three sets of the measured data ($t = 5, 6, 7$ in Table 3.2). It has to be noted that the measured data represents a part load operation, thus, the resulting inner efficiency might be slightly underestimated for a full load operation. Similar calculations for the LPT provide also an inner efficiency of 0.894, as shown in Table 3.3.

3.1.2 State Variables T-s and h-s Diagrams

With the above described calculations, the thermodynamic state variables at points 1 to 16 are described as a function of temperature and pressure, e. g. specific entropy $s_i = s(p_i, T_i)$ and enthalpy $h_i = h(p_i, T_i)$. The reference state for specific entropy is based on the Third Law of Thermodynamics ($s = 0\ kJ/kg$ at T = 0 K and p = 1.01325 bar) while the reference state for enthalpy is based on the enthalpy of formation relative to the elements at $25°C$ ($h = 0\ kJ/kg$ at $T = 298.15\ K$ and $p = 1.01325\ bar$). Table 3.4 lists the calculated thermodynamic variables for the Huntorf process, whilst Figures 3.2 and 3.3 show the Huntorf open cycle circuit as T-s and h-s-diagrams, respectively. The T-s-diagram (Fig. 3.2) displays additionally the reversible Huntorf process ($\eta_s = 1$), that is shown as overlay plot in grey to illustrate the effect of the irreversibilities. Fig. 3.4 displays the pressure-specific volume diagram of the Huntorf CAES process for a high cavern pressure of 68 bar.

Exergy values as combined property of the state and its surrounding are calculated with Eq.2.4 where ambient air conditions (marked with subscript ∞) correspond to the state point $i = 1$ in Table 3.4, hence $T_\infty = T_1 = 283\ K$, $h_\infty = h_1 = -15.5\ kJ/kg$ and $s_\infty = s_1 = 6808\ kJ/kgK$.

Compressibility factor. The compressibility factor ($z = p \cdot v / R_{air} \cdot T$) of an ideal gas is one. By calculating the compressibility factor with the calculated state variables of air treated as real gas, the deviation from ideal gas can be estimated for every state point i, see Fig. 3.5. The state points 8 and 13 with high pressure and high air temperatures have the highest z factors. Calculation with ideal gas law ($z = 1$) would cause an error of around 2 %.

3.1.3 Technical Work and Heat

Based on the above presented thermodynamic state variables, the enthalpy difference of each process stage can be determined. For the compression (C) and expansion (T) the enthalpy difference corresponds to the specific technical work (w_t). For the inter- and after-cooler, it corresponds to the transferred heat q_{loss} whilst for the burner to q_{fuel}. Thus, for 68 bar cavern

i	Pressure (p) in bar	Temperature (T) in K	Specific: Entropy (s) in J/kg-K	Enthalpy (h) in kJ/kg	Exergy (ex) in kJ/kg
1	1.013	283.0	6808	-15.5	0
2	**6.0**	503.0	6879	207.6	202.8
3	5.99	**308.0**	6380	8.6	145.2
4	13.48	418.3	6456	119.9	235.1
5	13.45	**308.0**	6143	7.1	210.7
6	30.27	413.1	6205	113.0	299.0
7	30.20	**309.0**	5904	4.7	276.0
8	**68.0**	431.8	6009	129.6	371.1
9	67.20	**322.0**	5698	11.9	341.5
10	67.20	322.0	5698	11.9	341.1
11	63.20	322.0	5718	12.6	336.6
12	42.00	318.6	5835	12.6	303.5
13	**41.3**	**763.0**	6765	484.5	512.3
14	13.50	585.6	6804	293.5	310.1
15	**12.8**	**1218.0**	7634	1002.0	783.6
16	1.013	699.8	7738	414.9	167.1

TABLE 3.4: Thermodynamic state variables and specific exergy of enthalpy of Huntorf CAES including **design data** in boldface print (for cavern pressure 68 bar)

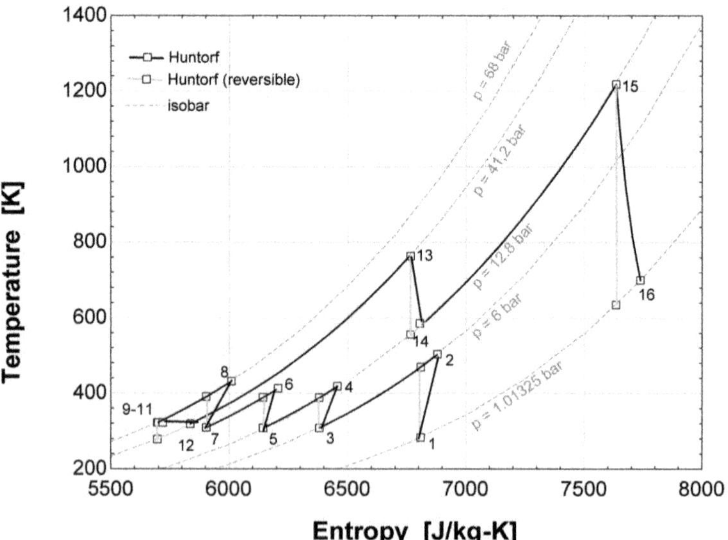

FIGURE 3.2: T-s diagram of Huntorf CAES: polytropic (black) and reversible (grey)

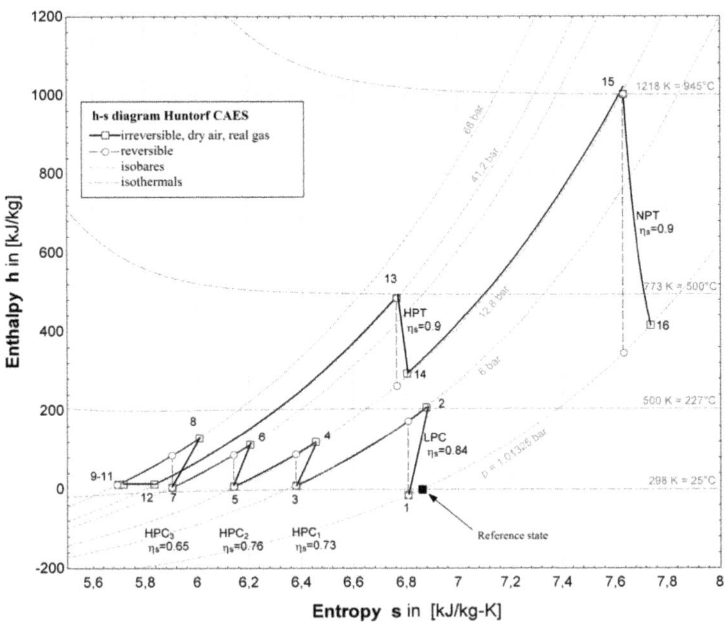

FIGURE 3.3: h-s diagram of Huntorf CAES

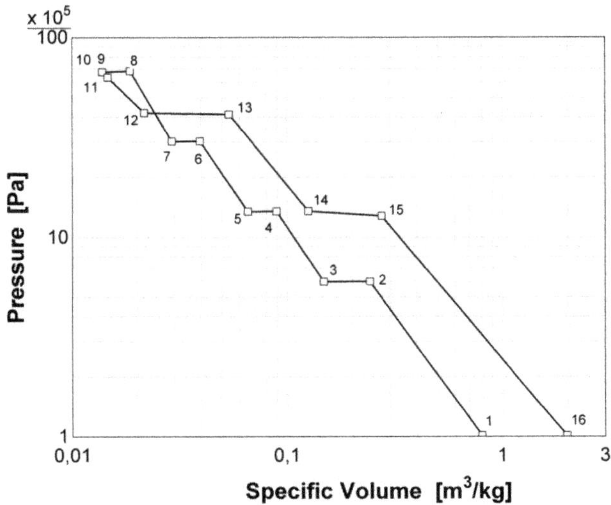

FIGURE 3.4: p-v diagram of Huntorf CAES

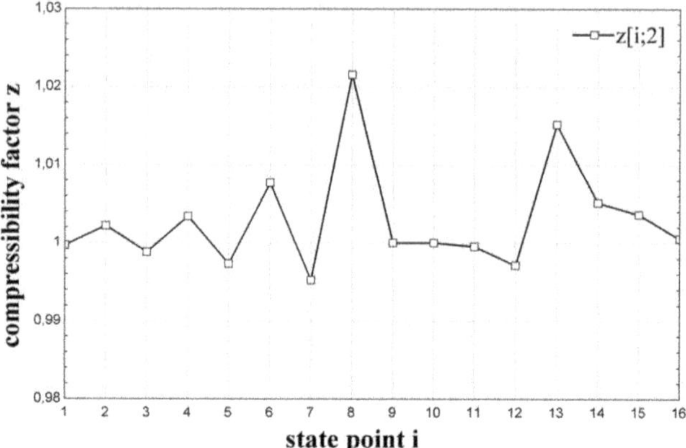

FIGURE 3.5: Deviation from ideal gas behavior

pressure the technical work of the low pressure compressor (LPC) amounts to $(h_2 - h_1) = 223\ kJ/kg$; for the high pressure compressor (HPC) a figure of $(h_4 - h_3 + h_6 - h_5 + h_8 - h_7) = 342\ kJ/kg$ is applicable. The heat removed after the LPC amounts to $(h_3 - h_2) = -199\ kJ/kg$ while the heat removed after the HPC is $(h_5 - h_4 + h_7 - h_6 + h_9 - h_8) = -339\ kJ/kg$. Both the work and heat figures are summarized in Table 3.5 and compared with values (given in brackets) corresponding to reversible processes ($\eta_s = 1$). Since the technical work for the compression increases with the cavern pressure, the lower (46 bar) and upper limit (68 bar) of the cavern pressure at full load operation are used as a parameter in Table 3.5. The comparison shows, that the technical work (w_t) of the polytropic and reversible processes differ significantly. Furthermore, it can be seen that a considerable part of the technical work to compress the air is dissipated in the inter- and after-cooling, hence lost energy (q_{loss}). The effect of throttling is not negligible as shows the comparison of the technical work for the $46\ bar$ and $68\ bar$ cavern pressure. During discharge, in a full load operation (cavern pressure in the 46 to $68\ bar$ range), the pressure has no effect on the technical work of the turbines since the turbine inlet pressure remains constant at $41.3\ bar$ due to the throttling.

In the HP burner, natural gas is burned to heat up the compressed air to a temperature of $T_{13} = 763\ K$. Thus, the heat added amounts to $h_{13} - h_{12} = q_{fuel}(HPT) = 472\ kJ/kg_{air}$ and a work of $h_{14} - h_{13} = -191\ kJ/kg$ is obtained from this turbine unit. For the LP expansion $h_{15} - h_{14} = q_{fuel}(LPT) = 709\ kJ/kg_{air}$ is supplied and a technical work of $-587\ kJ/kg$ is obtained. The exhaust gas has a temperature of approximately $T_{16} = 700\ K$, hence the exhaust gas enthalpy is $h_{16} = 415\ kJ/kg$.

3.1.4 Mechanical Efficiency

During charging the cavern with 108 kg/s air flow rate at 68 bar cavern pressure, it has been measured that the LPC and HPC units take from the electrical grid a power of 27 MW and 41 MW, respectively [Krüger, 27.07.2015]. On re-powering, at 455 kg/s air flow rate, the turbines (HPT+LPT) deliver 321 MW to the grid. Thus, the data allows for estimating the

Specific Technical Work (w_t) and Heat Exchanged (q) for polytropic (and reversible) process in [kJ/kg]			
Parameter:	Cavern Pressure	46 bar	68 bar
Charge	$w_{t,LPC}$	223 (188)	223 (188)
	$w_{t,HPC}$	281 (199)	342 (243)
	$w_{t,C}$ (total)	504 (387)	565 (431)
	q_{loss} (cooling A to D)	-473 (-357)	-538 (-404)
Discharge	$w_{t,HPT}$	-191 (-222)	
	$w_{t,LPT}$	-587 (-657)	
	$w_{t,T}$ (total)	-778 (-879)	
	q_{fuel} (burner)	1180 (1255)	

TABLE 3.5: Specific technical work (w_t) and heat exchanged (q) for the Huntorf plant for the minimum (46 bar) and maximum (68 bar) cavern pressure; values in brackets correspond to the reversible processes ($\eta_s = 1$)

mechanical efficiency (η_{mech}) for compression:

$$\eta_{mech,C} = \frac{P_{thermod.}}{P_{el.}} \tag{3.5}$$

and for expansion:

$$\eta_{mech,T} = \frac{P_{el.}}{P_{thermod.}} \tag{3.6}$$

where $P_{el.}$ stands for the electrical power taken from or delivered to the grid while $P_{thermod.}$ is the thermodynamic power of the compression/expansion part of the cycle that is simply the product of the air mass flow rate and the specific technical work listed in Table 3.5. Such calculated mechanical efficiency (η_{mech}) includes all mechanical losses of the compressor/turbine train, the shaft, the clutches and the motor/generator unit; the values are presented in Table 3.6. The mechanical efficiency of the whole turbine train is estimated to be 0.91 (see Table 3.6). Assuming that the overall mechanical efficiency of the turbine train applies to both turbine units, a 79 MW power is delivered to the grid by the HPT, while 243 MW by the LPT (322 MW in total).

Process Unit	Electrical Power $P_{el.}$ in MW	Thermodynamic Power $P_{thermod.}$ in MW	Mechanical Efficiency η_{mech}
LPC	27	24	0.89
HPC	41	37	0.90
Compression	68	61	0.90
HPT	n.a.	87	n.a.
LPT	n.a.	267	n.a.
Expansion	321	355	0.91

TABLE 3.6: Electrical and thermodynamic power as well as mechanical efficiencies for the Huntorf CAES plant (68 bar cavern pressure)

Similar considerations apply to the enthalpy added to the fluid stream during combustion ($\dot{Q}_{thermod.}$) in comparison with the fuel enthalpy ($\dot{Q}_{fuel}$). Heat losses from the burner to the surroundings can be taken into account by the burner efficiency:

$$\eta_{burner} = \frac{\dot{Q}_{thermod.}}{\dot{Q}_{fuel}} \tag{3.7}$$

It is assumed that heat losses in the burner are negligible and the burner efficiency is one.

The Huntorf CAES plant uses natural gas of type "L" according to the specification of DVGW [DVGW Deutsche Vereinigung des Gas- und Wasserfaches e. V., Mai 2008]. For further estimates the Lower Calorific Value (LCV) of $8.861\ kWh/m^3$ ($41\ MJ/kg$) is used.

3.1.5 Energy Storage Efficiencies

To calculate the different efficiency values developed in Chapter 2 from the above estimated specific technical work (w_t), one has to estimate the overall energy taken from the grid through $W_{el,M} = \frac{W_C}{\eta_{mech}} = \frac{\dot{m}_c}{\eta_{mech}} \cdot \int_0^{t_c} w_{t,C}(t)dt$, for constant mass flow rate $\dot{m}_c$ and with t_c being the duration of the charging cycle. The specific technical work of compression, in the full load pressure range, increases approximately linearly from a minimum value of $504.3\ kJ/kg$ at a cavern pressure of $46\ bar$ to a maximum value of $565.3\ kJ/kg$ at $68\ bar$ (see Table 3.5) during one full load charging period $t_c = 19.8\ hours$ (assuming an isothermal cavern). Since the air mass flow ($\dot{m}_c$) is constant at $108\ kg/s$ the total compression work $W_C = \dot{m}_c \cdot \int_0^{t_c} w_{t,C}(t)dt$ amounts to $1144\ MWh$. With a mechanical efficiency $\eta_{mech} = 0.9$ (see Table 3.6), the energy taken from the grid amounts to $W_{el,M} = \frac{W_C}{\eta_{mech}} = 1271\ MWh$ (for an isothermal cavern). The calculation of $W_{el,G}$ is simpler since the specific technical work is constant ($\dot{m}_d(t) = const.$) throughout the full load discharging period (t_d), thus $W_{el,G} = \eta_{mech} \cdot \dot{m}_d \cdot \int_0^{t_d} w_{t,T}(t)dt = \eta_{mech} \cdot \dot{m}_d \cdot w_{t,T} \cdot t_d = 1514\ MWh$. Similarly $Q_{fuel} = \dot{m}_d \cdot q_{fuel} \cdot t_d = 2520\ MWh$ (for an isothermal cavern). When inserting the above values into Eq.(2.8), the air mass can be canceled and the CAES efficiency of $\eta_{caes} = 39.9\%$ is obtained. Crotogino [Crotogino, 2003] estimated the Huntorf plant efficiency (probably using the above definition) to be around 42 % and since then it is widely accepted as a reference value for the plant [Beck et al., 2013; Wolf, 2011; Nielsen, 2013; Ausfelder et al., 2015; Hartmann et al., 2012b]. If one disregards the $\eta_{rt1} = 119.1\ \%$ figure, the Huntorf CAES efficiencies span from 2.9 % to 66.4 % depending on the efficiency definition in use (Table 3.7). Under reversibility assumption ($\eta_s = 1$) the CAES efficiency value of the Huntorf plant would reach $\eta_{caes}(reversible) = 47.9\ \%$.

	η_{caes}/ξ_{caes} %	η_{th}/ξ_{th} %	η_{rt1}/ξ_{rt1} %	η_{rt2} %	η_{rt3} %	η_{rt4}/ξ_{rt4} %	hr_1 $\frac{kWh_{fuel}}{kWh_{electric}}$	hr_2
enthalpy	39.9	9.6	119.1	66.4	39.9	2.9	10.4	1.7
exergy	49.3	13.3	119.1			33.3		

TABLE 3.7: Comparison of the different efficiencies and heat rates for the Huntorf CAES

Since the heat dissipated during inter cooling (q_{loss}) is $473\ kJ/kg$ (for $46\ bar$ cavern pressure) and $538\ kJ/kg$ (for $68\ bar$ cavern pressure) (see Table 3.5) the overall heat loss ($Q_{loss} = \dot{m} \cdot \int_0^{t_c} q_{loss}(t)dt$) is estimated to be $Q_{loss} = 1081\ MWh$ for charging the caverns from $46\ bar$ to $68\ bar$ pressure over the $19.6\ h$ period. The energy content of the compressed and stored air is then $E_{air} = W_C - Q_{loss} = (1144 - 1081)\ MWh = 63\ MWh$. Thus, the compression conversion coefficient amounts to $\eta_{cc} = 0.029$, illustrated by Fig.3.7.

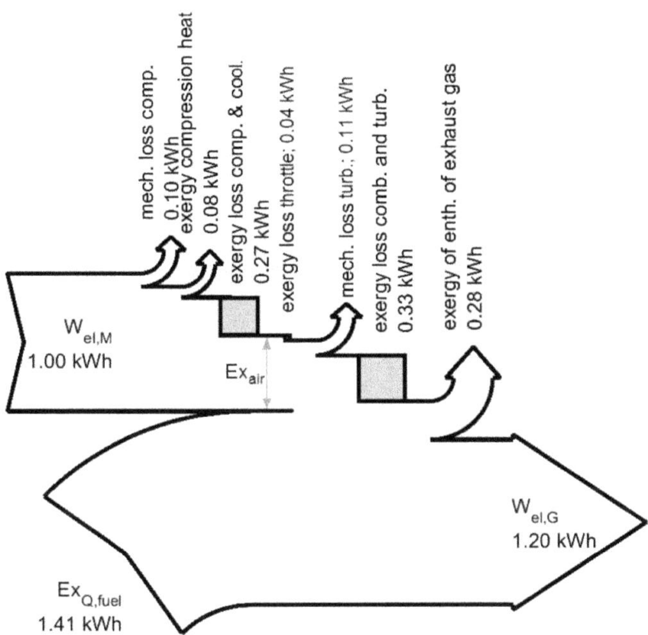

FIGURE 3.6: Sankey diagram of Huntorf CAES' exergy flows

On discharging, this compressed air energy (63 MWh) as well as the fuel energy (2520 MWh) enter the turbines and generate 322 MW power during 4.7 hours, which is an energy of 1513 MWh. Thus, the turbine conversion coefficient $\eta_{tc} = 0.586$ in the full load range. This turbine efficiency is considerably higher than for typical gas turbines. Hence, the 63 MWh energy stored in the compressed air is then converted in the turbines into 63 $MWh \cdot 0.586 = 37$ MWh work delivered back to the electrical grid. Fig. 3.7 shows the energy flow Sankey diagram based on the above values.

For the Huntorf CAES process the corresponding Sankey exergy flow diagram is presented in Fig.3.6. The exergetic efficiencies ξ often result in higher values then the corresponding enthalpy based efficiency, see Table 3.7. The exergetic CAES efficiency amounts to $\eta_{caes} =$ 49.3 % compared to the enthalpy based 39.9 % value. The round trip coefficient η_{rt1} remains equal because electric energy can be considered strictly exergy. The biggest difference occurs for the round trip efficiency 4 (η_{rt4} and ξ_{rt4}) value: While the enthalpy based value is very low ($\eta_{rt4} = 2.9$ %) the exergy based value is considerably higher with $\xi_{rt4} = 33.3$ %. This difference emphasizes the major difference in evaluation the energy content of compressed air at ambient conditions (E_{air}) and is well illustrated by the energy and exergy flow diagrams, Fig.3.7 and 3.6, respectively. The turbine conversion coefficient in exergetic terms amounts to $\xi_{tc} = 60.7$ % (compared to the enthalpy based $\eta_{tc} = 0.586$ value).

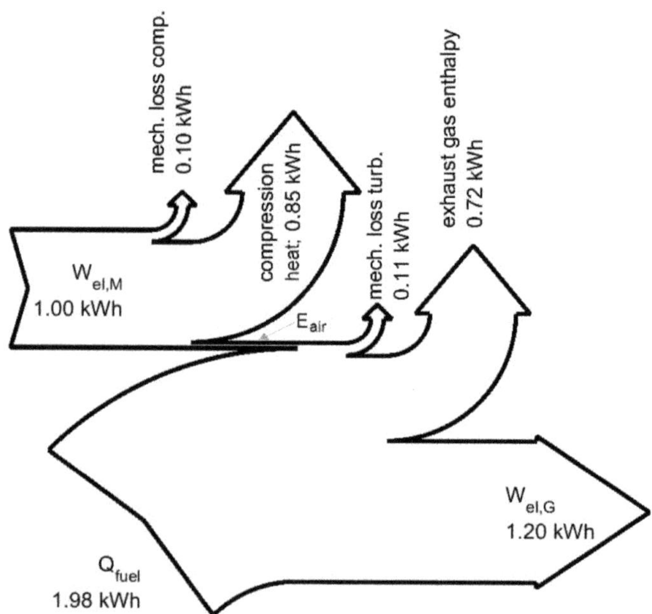

FIGURE 3.7: Sankey diagram of Huntorf CAES' energy flows

3.2 McIntosh Plant

In terms of design, the main difference between Huntorf and McIntosh plants is the additional exhaust enthalpy recuperation unit at McIntosh site. A process flow diagram is shown in Fig.3.8 with combustion air preheating at path 11-12 using the exhaust gas enthalpy (path 16-17). The thermodynamic data is summarized in Table 3.8.

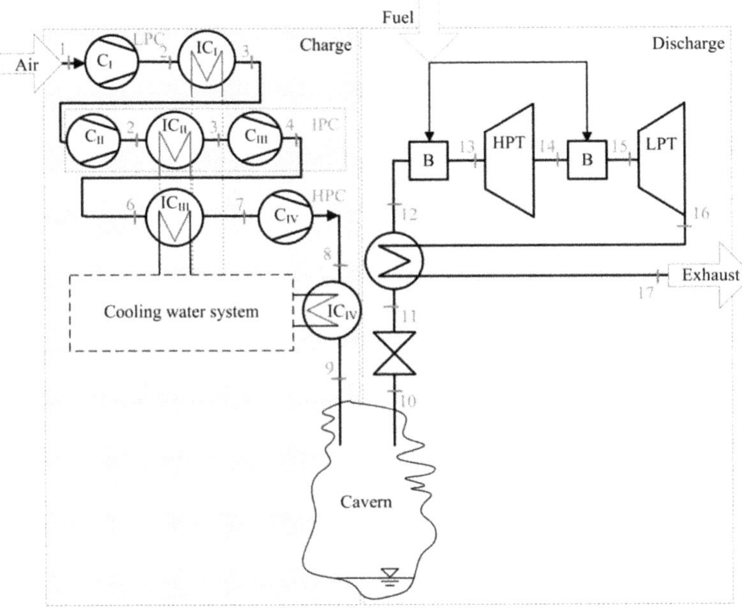

FIGURE 3.8: Process flow diagram of McIntosh CAES with exhaust gas enthalpy recuperation (adapted from [Pollak, 1994])

Charge. A compressor train that consists of one axial and two centrifugal compressors in series (shown in Fig. 3.8 as Low Pressure Compression (LPC), Intermediate Pressure Compression (IPC) and High Pressure Compression (HPC)) with a total electrical power of approximately 49 MW is used for charging [Pollak, 1994]. The overall pressure ratio (p_8/p_1) ranges from 50:1 to 75:1 [Pollak, 1994]. Cooling is applied as shown in Fig. 3.8. The inter-staged pressures of all compression stages (C_I to C_{IV}) for McIntosh plant are known [Pollak, 1994]. Pressure losses are estimated using the same procedures as for Huntorf. The mass-flow rate in charging mode is 89.4 kg/s [Pollak, 1994]. For the calculation the same inter- and after-cooling temperatures as in Huntorf are used.

Cavern. The air is stored inside a solution-mined salt cavern with a volume of approximately $538,000\ m^3$ [Pollak, 1994; Wolf, 2011]. The maximum allowable pressure inside the cavern is 77.9 bar [Nakhamkin et al., 1989].

i	p in bar	T in K	s in J/kg-K	h in kJ/kg	ex in kJ/kg
1	1.013	283	6808	-15.5	0.0
2	**(3.9-)4.3**	442	6842	144.8	150.6
3	4.29	**308**	6477	9.0	118.2
4	**(10.1-)11.7**	429	6523	131.4	227.3
5	11.7	**308**	6184	7.4	199.4
6	**(22.2-)27.7**	415	6237	115.4	292.3
7	27.67	**309**	5931	5.2	268.7
8	**(50-)75.3**	435	5985	132.3	380.4
9	74.4	**322**	5665	10.7	349.5
10	74.4	322	5665	10.7	349.5
11	45.5	318	5806	10.7	309.7
12	45.4	556	6396	261.2	393.2
13	**44.8**	**811**	6808	537.6	552.8
14	15.2	639	6861	349.4	349.8
15	**14.7**	**1144**	7520	915.2	728.8
16	1.16	654	7625	365.4	149.5
17	1.01	**330**	6963	31.9	3.5
Flow Rate and LCV					
	air in charge mode				**89.4 kg/s**
	air in discharge mode				**154.2 kg/s**
	fuel (natural gas)				n.a.
	LCV (natural gas)				n.a.

TABLE 3.8: Thermodynamic data of the McIntosh process including the **design data** in boldface print [Pollak, 1994] for a maximum pressure of 75.3 bar.

Discharge. During discharge a 154.2 kg/s compressed air flow leaves the cavern and enters directly the throttling process; in contrast to Huntorf no filter is installed [Pollak, 1994]. The compressed air then enters a heat exchanger where it is preheated using exhaust gas enthalpy. To determine the temperature T_{12} of the preheated air leaving the recuperator, it is assumed that 75 % of the exhaust gas enthalpy is recovered (in accordance with the design value [Pollak, 1994]). The pressure loss inside the recuperator for both paths 11-12 and 16-17 is estimated to be $< 0.2\ bar$. The temperature of the pre-heated compressed air is then further increased through combustion of natural gas before expanding in the HP and LP turbines.

3.2.1 Inner Efficiency

For Huntorf the inner efficiencies (η_s) of compression have been calculated using measured data whereas for McIntosh the data is given in the plant documentation [Pollak, 1994]. Since the design values of inlet and outlet temperatures and pressures for both turbines are given too, the inner efficiencies of the expansions can be calculated as for Huntorf plant. The resulting values are listed in Table 3.9 in comparison with the Huntorf inner efficiencies.

McIntosh's inner efficiencies η_s for the compressor units are 4 to 17 percent points higher than those of the Huntorf process units. On the other hand, the Huntorf turbine train has a higher inner efficiency than the McIntosh ones (1 to 4 percent points).

	Process Unit	Inner Efficiency η_s Huntorf	McIntosh
Charge	C_I	0.844	0.907
	C_{II}	0.726	0.840
	C_{III}	0.764	0.801
	C_{IV}	0.653	0.819
Discharge	HPT	0.894	0.851
	LPT	0.894	0.882

TABLE 3.9: Inner efficiencies of the compressor and turbine units of McIntosh plant (based on data sheet specifications [Pollak, 1994]) compared to Huntorf (calculated from measured data [Krüger, 27.07.2015]) (C = compressor; T = turbine)

3.2.2 State Variables T-s and h-s Diagrams

The T-s and h-s diagrams of the McIntosh process are shown in Figures 3.9 and 3.10, respectively. The isobars correspond to the values of the Huntorf process to facilitate a direct comparison. The pressure loss inside the exhaust gas enthalpy recuperator is visible; the change of state 16-17 is not exactly isobaric. This also applies to other heat exchange processes depicted, but as the pressure losses are very small indeed, this is hardly visible in the diagrams.

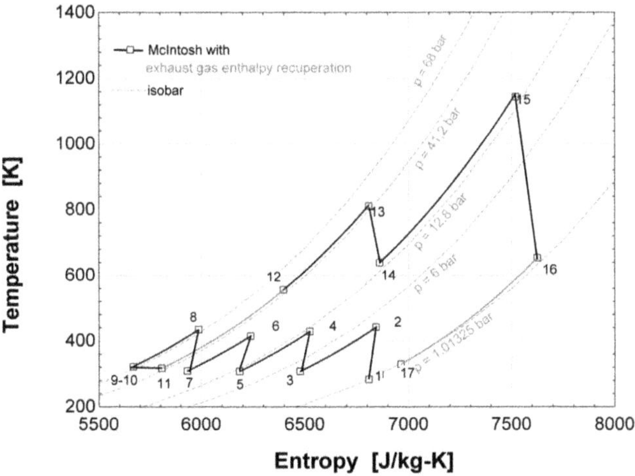

FIGURE 3.9: T-s diagram of McIntosh CAES with exhaust gas enthalpy recuperation

3.2.3 Technical Work and Heat

The specific technical work of McIntosh's process units, as well as heat removed or added, are shown in Table 3.10. The cavern pressure is used as a parameter to show the effect of throttling. Furthermore, it can be seen that most of the enthalpy added to the air

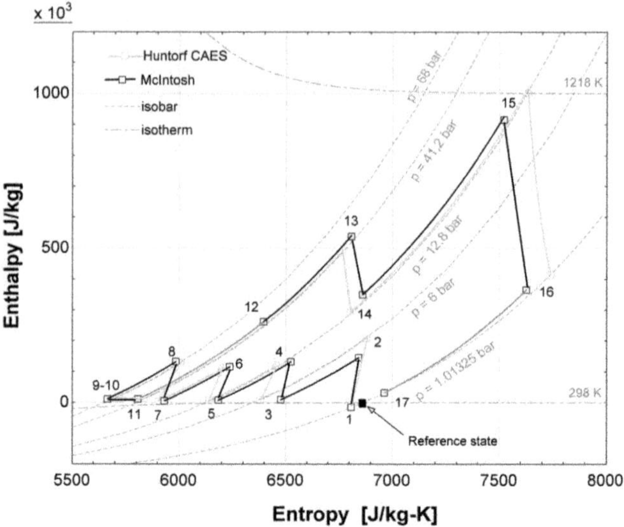

FIGURE 3.10: h-s diagram of McIntosh CAES (black) in comparison with Huntorf (grey)

stream during compression ($w_{t,C}(75.3\ bar) = 518\ kJ/kg$) is dissipated in the cooling process ($q_{loss}(75.3\ bar) = -419\ kJ/kg$) making the air compression a rather inefficient process with an efficiency of $\eta_{cc} = 0.054$. Table 3.10 also shows the enthalpy rate that is recuperated from the exhaust gas ($250\ kJ/kg$).

3.2.4 Mechanical Efficiency

The mechanical efficiencies of McIntosh's process units are calculated in the same way as for Huntorf (see Section 2.1.4), see Table 3.11, and they are slightly higher than Huntorf's values (compare with Table 3.6).

Cavern size and operation duration. The McIntosh solution mined salt cavern has a total volume of approximately $538,000\ m^3$. The discharge-charge mass flow ratio is 1.8 ($154.2\ kg/s : 85.4\ kg/s$) which is equal to the discharge-charge duration ratio of $49.8\ h : 27.6\ h$ for isothermal or $32.2\ h : 17.8\ h$ for adiabatic operation of the cavern. A realistic representation of the cavern considering heat transfer and dynamic effects is developed by Nakhamkin et al. [1989] estimating 41 hours of charging and 26 hours of discharging.

3.2.5 Energy Storage Efficiency

The efficiency values of the McIntosh CAES plant are calculated in the same way as those of the Huntorf plant (see Section 2.1.6) and summarized in Table 3.12.

The exergetic CAES efficiency values amounts to $\xi_{caes} = 62.9\ \%$, thermal exergetic efficiency is $\xi_{th} = 30.0\ \%$. Calculated with electric energy output to input the round trip 1 values are equal $\xi_{rt1} = \eta_{rt1} = 136.1\ \%$. The largest difference between enthalpy based and exergetic

Specific Technical Work (w_t) and Heat (q) in [kJ/kg]			
Parameter:	Cavern Pressure	50.0 bar	75.3 bar
Charge	$w_{t,LPC}$	147	160
	$w_{t,IPC}$	213	230
	$w_{t,HPC}$	100	127
	$w_{t,C}$ (total)	460	518
	q_{loss} (cooling A to D)	-430	-491
Discharge	$w_{t,HPT}$	-188	
	$w_{t,LPT}$	-550	
	$w_{t,T}$ (total)	-738	
	q (exhaust gas enthalpy recuperation)	250	
	q_{fuel} (burner)	842	

TABLE 3.10: Specific technical work and heat of the process units in McIntosh, calculated with inner efficiency values according to data sheet specification [Pollak, 1994]

Process Unit	Electrical Power $P_{el.}$ in MW	Thermodynamic Power $P_{thermod.}$ in MW	Mechanical Efficiency η_{mech}
LPC	n.a.	14.3	n.a.
IPC	n.a.	20.6	n.a.
HPC	n.a.	11.4	n.a.
Compression	49	46.3	0.93
HPT	26.5	29.0	0.91
LPT	87.4	84.8	0.97
Expansion	110	113.8	0.97

TABLE 3.11: Comparison of electrical and thermodynamic power as well as mechanical efficiency for the McIntosh CAES plant

	η_{caes} %	η_{th} %	η_{rt1} %	η_{rt2} %	η_{rt3} %	η_{rt4} %	hr_1 $\frac{kWh_{fuel}}{kWh_{electric}}$	hr_2
McIntosh	52.3	22.6	136.1	83.0	72.1	4.5	4.4	1.2

TABLE 3.12: Comparison of the efficiency values of McIntosh CAES

efficiency occurs at the round trip 4 value: $\xi_{rt4} = 47.8\ \%$, which is ten times larger than the enthalpy based value ($\eta_{rt4} = 4.5\ \%$). Again, the fundamentally different evaluation of the energy content of compressed air causes this difference. While in enthalpy based considerations the energy in the storage state (E_{air}, see Fig.3.11) is low and the corresponding compression conversion coefficient $\eta_{cc} = 5.4\ \%$, the exergetic energy content of air is higher (Es_{air}, see Fig.3.12) leading to $\xi_{cc} = 62.7\ \%$. The turbine conversion coefficients are $\eta_{tc} = 82.4\ \%$ and $\xi_{tc} = 76.2\ \%$.

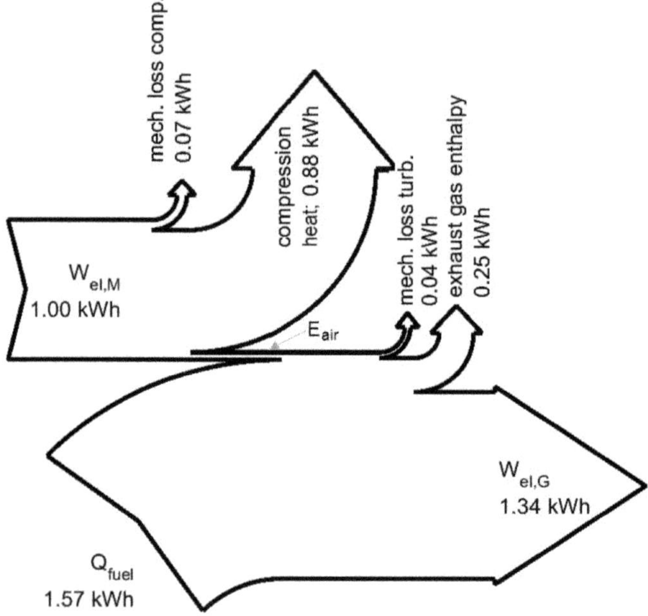

FIGURE 3.11: Sankey diagram of McIntosh CAES' energy flows

3.3 Next Generation CAES

The advanced CAES concepts considered in this thesis use adiabatic, isobaric or quasi-isothermal processes to enable either a more efficient or a fuel-free operation. While considering these advanced concepts the inner efficiencies of compression and expansion have to be estimated. Therefore, the average of Huntorf's and McIntosh's inner efficiencies are rounded to one decimal place and used as a "good guess" value for the advanced CAES concepts. The values are presented in Table 3.13.

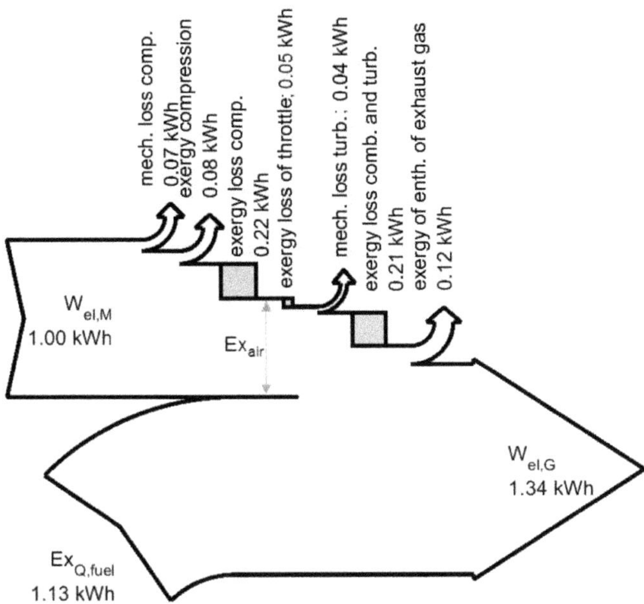

FIGURE 3.12: Sankey diagram of McIntosh CAES' exergy flows

	Process Unit	Inner Efficiency η_s		
		Huntorf	McIntosh	Advanced
Charge	C_I	0.844	0.907	0.9
	C_{II}	0.726	0.840	0.8
	C_{III}	0.764	0.801	0.8
	C_{IV}	0.653	0.819	0.8
Discharge	HPT	0.894	0.851	0.9
	LPT	0.894	0.882	0.9

TABLE 3.13: Inner efficiencies of the compressor and turbine units of Huntorf, McIntosh and the advanced concepts (C_n - compressor stage n; HPT - high pressure turbine; LPT - low pressure turbine)

3.3.1 Adiabatic CAES (ADELE)

"ADELE"[2] is a German project aiming at the development of an adiabatic CAES (ACAES). At the beginning mainly driven by the German power supplier RWE, the project started in 2009 but the actual construction of a demonstration plant has been delayed several times, because some technical and economic challenges remain unsolved [Moser et al., 2012; Zunft, 2015].

A process flow diagram of ADELE ACAES is shown in Fig. 3.13 and the thermodynamic data is summarized in Table 3.14. The basic idea of ACAES is to avoid the use of fuels by storing the heat removed during compression to reuse it during expansion. Thus, the word 'adiabatic' signifies that no heat, in the form of fuel, is added to the process making it a strictly electrical energy storage facility. Instead of fuel addition, a heat storage is used to store the heat generated during compression and reuse it during expansion, as shown in Fig. 3.13. Therefore, the outlet temperature of the compression (T_4) has to be relatively high to reach a sufficiently high temperature level for the expansion process downstream so as to avoid icing in the turbines. Thus, little or no inter-cooling during compression is desired.

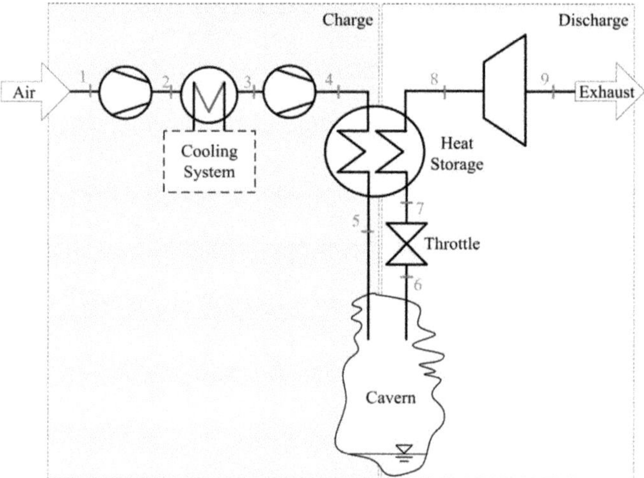

FIGURE 3.13: Process flow diagram of adiabatic CAES 'ADELE'

The technical challenges are in developing a compressor that withstands high outlet temperatures and in designing a heat storage that has a small heat loss during long-term storage. Furthermore, the heat storage has to deliver, during the entire charging and discharging, constantly low (at the cavern inlet) and constantly high (at the turbine inlet) temperatures, respectively, despite the non-steady state operation. Finally, the development of a turbine withstanding high inlet pressures is another technical challenge. In ADELE project, General Electric (GE) is in charge of developing a high pressure, high temperature compressor as well as a high pressure turbine; Ed. Zueblin AG and DLR develop heat storage technical solutions [Moser et al., 2012; RWE Power AG, 2010; Zunft et al., 2012; Laing et al., 2013; Steinmann, 2014]. The envisaged parameters are a maximum temperature (T_4) of around 650°C (923 K) and a maximum pressure (p_4) of 70 bar at the compression unit outlet [Moser

[2] ADELE = Adiabater Druckluftspeicher fuer die Elektrizitaetsversorgung - Adiabatic Compressed Air Storage for the Electricity Supply System

i	Pressure (p) in bar	Temperature (T) in K	Specific Entropy (s) in J/kg-K	Specific Enthalpy (h) in kJ/kg
1	1.013	283	6808	-15
2	8.4	541	6858	247
3	8.4	470	6712	174
4	**70**	**923**	6824	665
5	69.8	*322*	5686	12
6	69.8	322	5686	12
7	*42.0*	318	5831	12
8	*41.3*	772	6775	494
9	1.01	325	6948	27

TABLE 3.14: Thermodynamic data of ADELE with **design data** [RWE Power AG, 2010] in bold print.

et al., 2012; Zunft et al., 2012]. An earlier publication on ADELE [RWE Power AG, 2010] suggested pressures up to 100 bar. Since only the maximum temperature and pressure are known, several assumptions are needed (see below).

Charge. It is assumed that the inter-staged pressure (p_2) is optimized so as to minimize the technical work for the compression, hence $p_2 = \sqrt{70}\ bar = 8.4\ bar$. Thus, T_2 can be calculated through Eq.(5) using the inner efficiency of 0.9 (see Table 3.13) to be 541 K. The outlet pressures of the inter-cooler p_3 is assumed to be the same as p_2 since the pressure losses in the heat exchanger (path 2-3) are small. The inter-cooling temperature T_3 is unknown, but can be estimated using Eq.(5) and the outlet temperature and pressure of the second compression stage (T_4 and p_4) which gives a value of $T_3 = 470\ K$. It is assumed that the maximum cavern temperature is the same as in Huntorf, so $T_5 = 322\ K$.

Discharge. For discharging no plant operation parameters have been published. Thus, a throttling pressure $p_7 = 42\ bar$ is chosen for a good comparability with Huntorf. If a turbine withstanding 60 bar pressure or higher is developed, p_7 pressure will be considerably higher to avoid throttling. In calculations of the outlet temperature of the heat storage (T_8), it is assumed that the same efficiency as for the McIntosh exhaust gas enthalpy recuperator of 75 % applies. It has to be noted that this may lead to an overestimation of the overall storage efficiency since the heat storage is being drained during the discharging process leading to smaller temperature differences and, thus, a lower outlet temperature, eventually resulting in a decrease of overall efficiency. The turbine inlet pressure is taken to be $p_8 = 41.3\ bar$ corresponding to the HPT of Huntorf.

T-s and h-s diagrams. The thermodynamic diagrams that correspond to the ACAES data listed in Table 3.14 are shown in the T-s diagram (Fig. 3.14) and the h-s diagram together with the Huntorf process in grey (Fig. 3.15). It can be seen that the maximum air temperatures of ADELE (923 K) are lower than Huntorf's (1218 K). The high temperatures of the Huntorf plant do occur during combustion, whereas the highest temperatures of ADELE are found at the compression, exceeding today's temperature range of turbo compressors.

Technical work. The specific technical work for the compression, at a cavern pressure of 70 bar, is $w_{t,C} = 753\ kJ/kg$ or at, a cavern pressure of 50 bar, $w_{t,C} = 654\ kJ/kg$. For

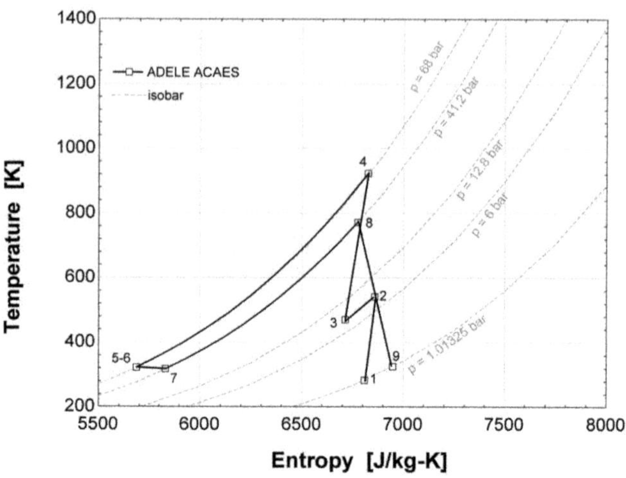

FIGURE 3.14: T-s diagram of adiabatic CAES 'ADELE'

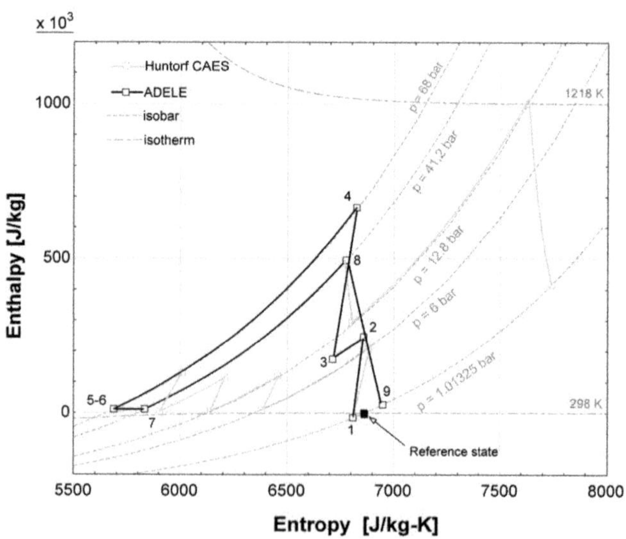

FIGURE 3.15: h-s diagram of the adiabatic CAES 'ADELE' (black) in comparison with Huntorf (grey)

the expansion a value of $w_{t,T} = -467\ kJ/kg$ is applicable. Thus, the net technical work (absolute values of expansion minus compression work) sums up to $w_{t,net} = -286\ kJ/kg$ or $-188\ kJ/kg$, respectively. The net work is negative, because ACAES is, in contrast to the other CAES configurations, a storage technology without fuel addition. Therefore, the output (technical work of the expansion) has to be lower than the input (technical work of the compression).

Power. The mass flow rate that is envisaged at ADELE project is unknown to the authors. For the calculation of power a mass flow rate of 100 kg/s during charging and discharging is assumed. The values for mechanical efficiency are taken to be the same as for Huntorf, $\eta_{mech,T/C} = 0.9$, see Table 3.6. Thus, for the compression a maximum power of 84 MW is taken from the grid while -42.5 MW are delivered back to the grid upon re-powering.

Energy storage efficiency. Obviously thermal efficiency, as defined by Eq.(2.9), is not applicable to ACAES since no fuel is added to the process ($Q_{fuel} = 0$), hence the heat rate equals zero. Other formulas (Eq.(2.11) to (2.13) and their exergetic counterparts) collapse to the same value 54.4 %, Table 3.15.

	η_{caes} %	η_{th} %	η_{rt1} %	η_{rt2} %	η_{rt3} %	η_{rt4} %	hr_1	hr_2
							$\frac{kWh_{fuel}}{kWh_{electric}}$	
ADELE	54.4	n.a.	54.4	54.4	54.4	54.4	0	0

TABLE 3.15: Comparison of the efficiency values of ADELE ACAES

This steady state calculation contains the simplifying assumption that the heat storage has a constant efficiency of 75 % (in analogy to the exhaust gas enthalpy recuperator in McIntosh) without taking into account transient effects associated with the heat transfer. This estimate is in line with the 60 % figure of Hartmann et al. [2012a] and 50 to 60 % figures of Pickard, Hansing, and Shen [2009].

3.3.2 Isobaric CAES (ISACOAST)

"ISACOAST-CC"[3] is an isobaric CAES concept, see Fig. 3.16 and Table 3.16, developed at the Braunschweig Technical University [Nielsen, 2013]. It combines several advanced features of CAES as: isobaric air storage in a salt cavern with shuttle pond, a heat storage unit comparable to the one used for adiabatic concepts, and a combined steam turbine cycle. Due to the isobaric air storage a turbine inlet pressure can be kept constant without throttling. Furthermore, since no off the shelf turbo machinery for CAES exists, Nielsen [2013] suggests to use a modified Alstom 'GT26' gas turbine in order to reduce the research and development efforts. Thus, in addition to the GT26-compressor that reaches up to 34 bar, an additional compressor is required to reach the envisaged 47 bar cavern pressure.

Charge. For charging a two stage compression without inter cooling is proposed; path 1-2 with $p_2 = 33.5\ bar$ and path 3-4 with $p_3 = 45.6\ bar$. The compressed air is cooled in the heat storage (path 3-4) to the permissible cavern temperature of $T_4 = 322\ K$. The cavern is kept under a constant pressure of $45\ bar$ using the hydro static pressure of an above ground shuttle pond filled with brine with a head of 340 m [Nielsen, 2013].

[3]ISACOAST-CC = Isobaric Adiabatic Compressed Air Energy Storage-Combined Cycle

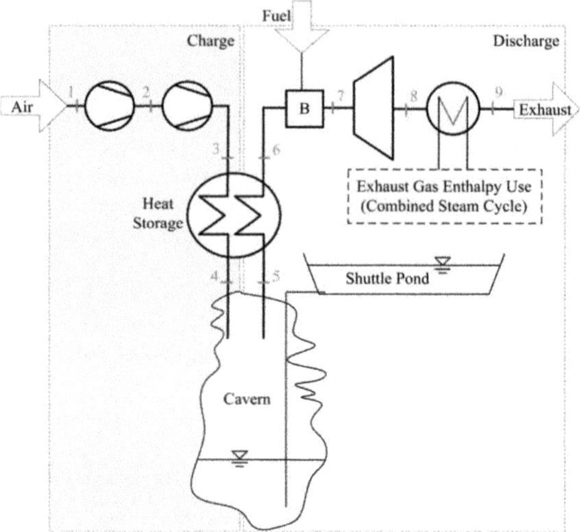

FIGURE 3.16: Process flow diagram of isobaric CAES 'ISACOAST-CC'

i	Pressure (p) in bar	Temperature (T) in K	Specific Entropy (s) in J/kg-K	Specific Enthalpy (h) in kJ/kg
1	1.013	288	6808	-15
2	**33.5**	801	6878	526
3	**(43-)45.6**	884	6899	619
4	45.4	**322**	5822	16
5	43.4	322	5836	16
6	43.0	**(563-)920**	6961	660
7	42.3	**1707**	7697	1595
8	1.04	789	7861	512
9	1.013	**360**	7050	62

TABLE 3.16: Thermodynamic data of ISACOAST with design parameter [Nielsen, 2013; Nielsen et al., 22.-23.03.2012] in bold face

Discharge. In discharging, the stored compressed air leaves the cavern passing through the heat storage for preheating (path 5-6). During discharging of the air cavern the heat storage is also being drained; the temperature within the heat storage decreases and so does the temperature of the preheated compressed air which is estimated to vary in the $T_6 = 920\ K$ to $T_{6'} = 563\ K$ range [Nielsen, 2013]. The preheated air is then mixed with a fuel and burned in the burner (6-7 or 6′-7) to boost the temperature to $T_7 = 1707\ K$ before entering the gas turbine (7-8). After the expansion, the enthalpy of the hot turbine exhaust gas ($T_8 = 789\ K$) is used (path 8-9) to run a steam turbine to provide electricity. This additional steam turbine is neglected in the thermodynamic considerations presented in this thesis. The temperature, pressure, specific entropy and enthalpy listed in Table 3.16 are calculated in the same way as described for the previous plant configurations. The inner efficiencies are given in Table 3.13 (0.9 for LP and 0.8 for HP processes). These values are slightly different to those used by Nielsen [2013] who used 0.861 for compression and 0.876 for expansion.

T-s and h-s diagrams. In the ISACOAST-CC process the temperature of the compressed air at the turbine inlet is 1707 K which is much higher than in the previous CAES processes considered, as illustrated by the T-s and h-s diagrams (Figures 3.18 and 3.17, respectively).

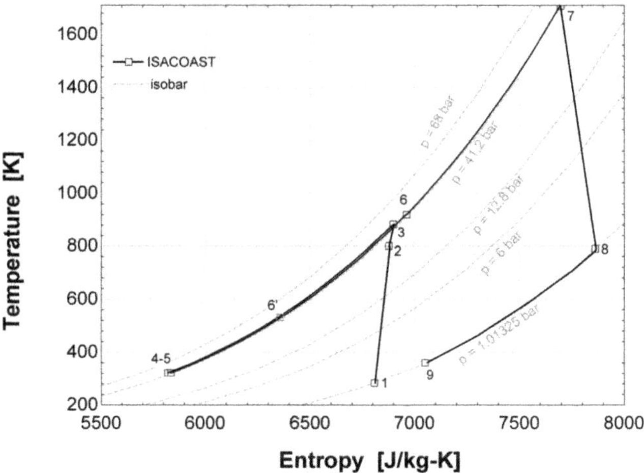

FIGURE 3.17: T-s diagram of isobaric CAES 'ISACOAST'

Technical work. The specific technical work of the compression unit ($w_{t,C}$) amounts to $634\ kJ/kg$. Since the air storage is kept at an almost constant pressure, this value is constant over the entire charging period. The turbine delivers a specific work of $-1082\ kJ/kg$, which is considerably higher than in any of the other concepts due to the extremely high turbine inlet temperatures. Thus, theoretically, a net work of $448\ kJ/kg$ can be obtained, or even more if one considers also the combined steam turbine. The heat storage depletion from state point 6 to 6′ during discharging has a major impact on fuel consumption. The fuel demand at a full and an empty heat storage is 924 and $1326\ kJ/kg_{Air}$, respectively.

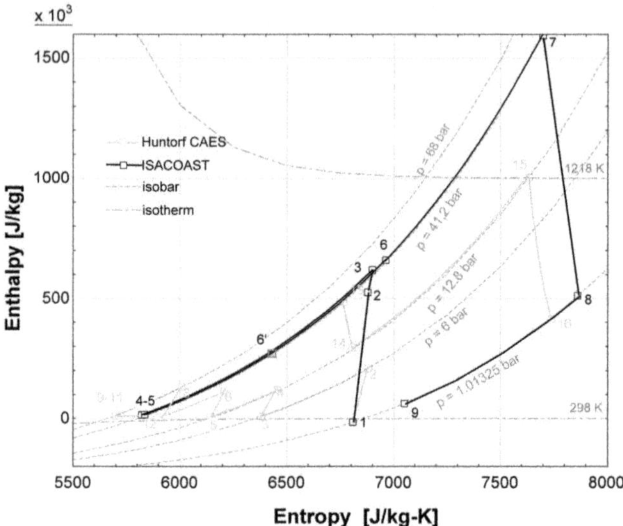

FIGURE 3.18: h-s diagram of isobaric CAES 'ISACOAST'

Energy storage efficiency. Nielsen [2013], using 0.861 inner efficiency for compression and 0.876 for expansion, estimates that the ISACOAST CAES efficiency to be 67 %. This figure is confirmed by the calculations at hand, if the same conditions as in [Nielsen, 2013] are applied: $\eta_{caes} = 66.3\ \%$ is obtained, when the heat storage is fully loaded and mechanical losses are neglected. Yet, to enable a comparison with the previously presented concepts the same calculation method as presented above is followed and the inner efficiencies of Table 3.13, as well as the mechanical efficiencies of the Huntorf process given in Table 3.6, are used. Then, the CAES efficiency as defined by Eq.(2.8) is found to be $\eta_{caes} = 50.3\ \%$, see Table 3.17, which is considerably lower than previously quoted 66.3 % value. Table 3.17 shows the resulting efficiencies previously defined. They are illustrated by Fig.3.19.

	η_{caes} %	η_{th} %	η_{rt1} %	η_{rt2} %	η_{rt3} %	η_{rt4} %	hr_1 $\frac{kWh_{fuel}}{kWh_{electric}}$	hr_2
ISACOAST	50.3	22.3	139.7	81.7	68.6	3.4	4.5	1.3

TABLE 3.17: Comparison of the efficiency values of ISACOAST CAES process

The exergetic CAES efficiency amounts to $\xi_{caes} = 59.7\ \%$. Thermal efficiency based on exergy is $\xi_{th} = 29.7\ \%$. Again, round trip efficiency 1 is equal for both enthalpy and exergy based considerations. Round trip efficiency 4 shows the largest deviation due to the different evaluation of the storage state $\xi_{rt4} = 56.1\ \%$ (compared to the $\eta_{rt4} = 3.4$ value) with $\xi_{cc} = 90.0\ \%$ and $\xi_{tc} = 62.4\ \%$ (compared to $\eta_{cc} = 4.4\ \%$ and $\eta_{tc} = 84.6\ \%$). The exergetic values are illustrated by Fig.3.20. These values are however based on a process set-up with comparatively high temperatures that exceed today's state of the art which will be discussed in Section 3.4.

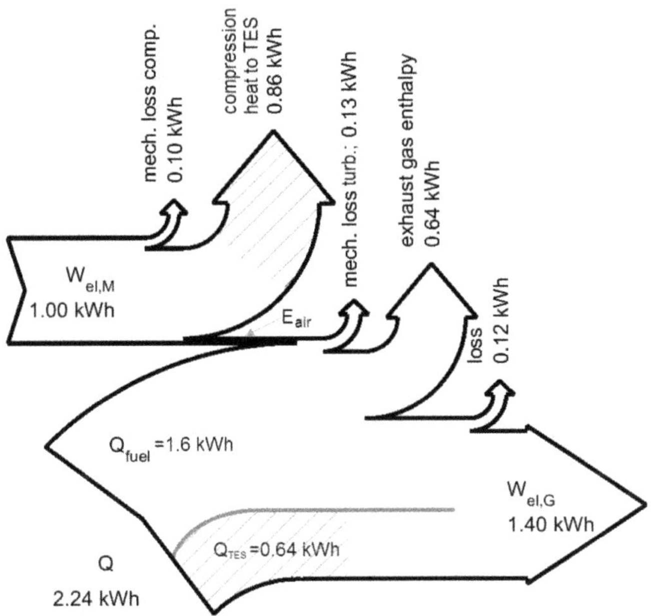

FIGURE 3.19: Sankey diagram of ISACOAST CAES' energy flows

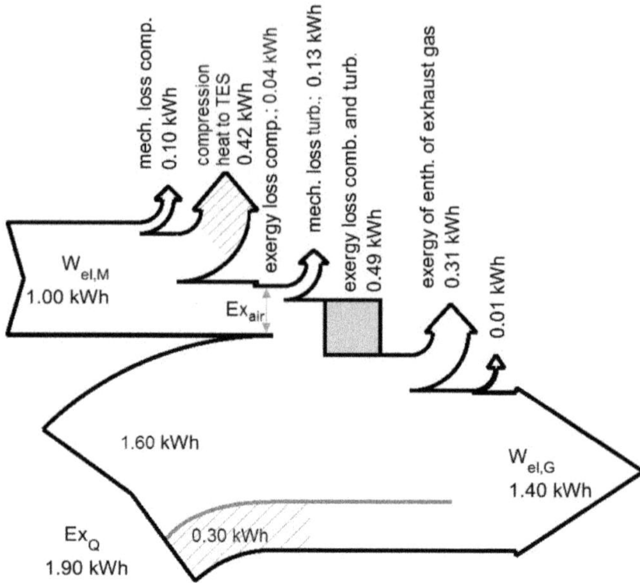

FIGURE 3.20: Sankey diagram of ISACOAST CAES' exergy flows

3.3.3 Quasi-Isothermal CAES (Sager Meer)

The "Sager Meer" concept [4] is based on the idea of CAES utilizing a quasi-isothermal compression [Oldhafer et al., 2014; Leithner, 20.01.2015]. The concept, see Fig. 3.21 and Table 3.18, has been developed at the Braunschweig Technical University together with Umwelttechnik & Ingenieure GmbH [Oldhafer et al., 2014; Leithner, 20.01.2015]. For this concept a compressor train, that is usually used in air separation with a built-in cooling, can be used [Oldhafer et al., 2014; Leithner, 20.01.2015]. Such a compression train is commercially available in units of up to 100 MW power. No fuel combustion is foreseen but instead a water heat storage is proposed. During expansion the air is heated to 370 K temperature in all expansion stages and over the entire discharging cycle. To achieve this, the system is connected to a local heat distribution grid.

Charge. The quasi-isothermal compression is implemented through six compression stages, with $\beta = \frac{p_{i+1}}{p_i} = 2.03$ compression ratio for each stage [Oldhafer et al., 2014; Leithner, 20.01.2015]. Since there is a small pressure loss (in-line with the calculation methods of the previously presented concepts) associated with each cooling, the inter-stage pressures are $p_2 = 2.1\ bar$, $p_4 = 4.1\ bar$, $p_6 = 8.2\ bar$, $p_8 = 16.5\ bar$, $p_{10} = 33.4\ bar$ and $p_{12} = 67.6\ bar$. After each compression stage the air is cooled to a temperature of $30°C$ ($T_3 = T_5 = T_7 = T_9 = T_{11} = 303\ K$). The air is stored at the maximum cavern temperature of $T_{13} = 322\ K$ (in analogy to the Huntorf concept). A salt cavern is suggested as storage place, but its non-isobaric

[4] Sager Meer lake is a site in Northern Germany for which the concept was developed.

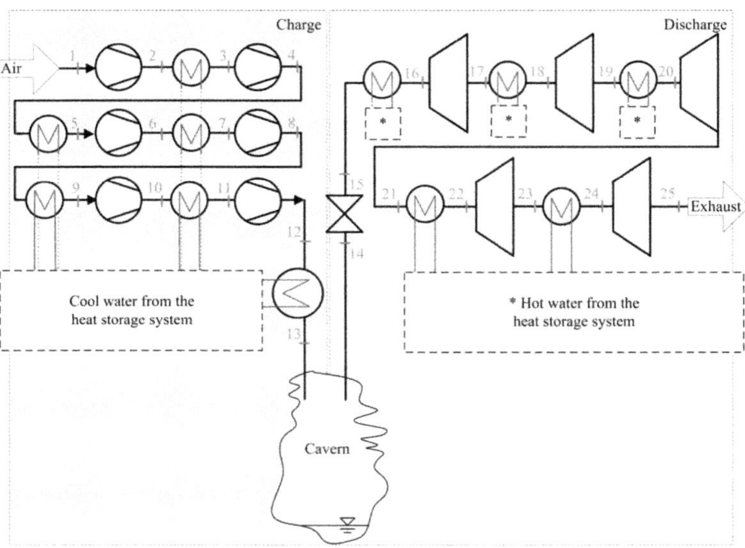

FIGURE 3.21: Process flow diagram of quasi-isothermal CAES 'Sager Meer'

i	Pressure (p) in bar	Temperature (T) in K	Specific Entropy (s) in J/kg-K	Specific Enthalpy (h) in kJ/kg
1	1.013	283	6808	-15
2	**2.1**	354	6828	55
3	2.0	303	6680	5
4	**4.1**	388	6725	90
5	4.0	**303**	6479	4
6	**8.2**	388	6524	90
7	8.1	**303**	6275	3
8	**16.5**	388	6320	89
9	16.4	**303**	6067	1
10	**33.4**	388	6112	87
11	33.3	**303**	5854	-2
12	**67.6**	388	5899	84
13	67.6	**322**	5696	12
14	63.6	322	5716	13
15	63.6	322	5716	13
16	63.6	**370**	5866	65
17	28.3	300	5892	-5
18	28.2	**370**	6113	69
19	12.5	301	6139	0
20	12.5	**370**	6353	71
21	5.5	301	6379	2
22	5.5	**370**	6591	72
23	2.4	301	6617	3
24	2.4	**370**	6831	72
25	1.1	301	6857	3

TABLE 3.18: Thermodynamic data of Sager Meer quasi-isothermal CAES concept

behavior is not taken into consideration in [Oldhafer et al., 2014; Leithner, 20.01.2015], so an isobaric storage (such as in the ISACOAST concept, see previous section) is assumed.

Discharge. A pressure loss of 4 bar during discharging is assumed, as measured at the Huntorf plant, hence $p_{15} = 63.6\ bar$ and $T_{15} = 322\ K$. The air then enters the first heat exchanger where it is heated to $T_{16} = 370\ K$. Due to the low turbine inlet temperature, several expansion stages are necessary to prevent icing in the turbines. There are five turbines expanding the compressed air to atmospheric pressure, thus, an expansion ratio of 2.25 is applicable, which is slightly lower than the value of 2.35 suggested in [Oldhafer et al., 2014; Leithner, 20.01.2015] caused by the estimated pressure losses. Thus, the inter-stage turbine pressures are $p_{16} = 63.6\ bar$, $p_{18} = 28.2\ bar$, $p_{20} = 12.5\ bar$, $p_{22} = 5.5\ bar$ and $p_{24} = 2.4\ bar$. The inlet temperature of each expansion stage is kept constant at $T_{16} = T_{18} = T_{20} = T_{22} = T_{24} = 97°C\ (370\ K)$, as shown in Table 3.18.

T-s and h-s diagrams. Fig. 3.22 and 3.23 show the T-s and h-s diagrams of the Sager Meer quasi-isothermal process, respectively. The overall process is turning anti-clockwise (energy consuming). The temperatures range from ambient to a maximum value of 388 K and are, thus, considerably lower than in other concepts considered in the comparison, well illustrated by Fig. 3.23 showing also the Huntorf process as overlay plot in grey.

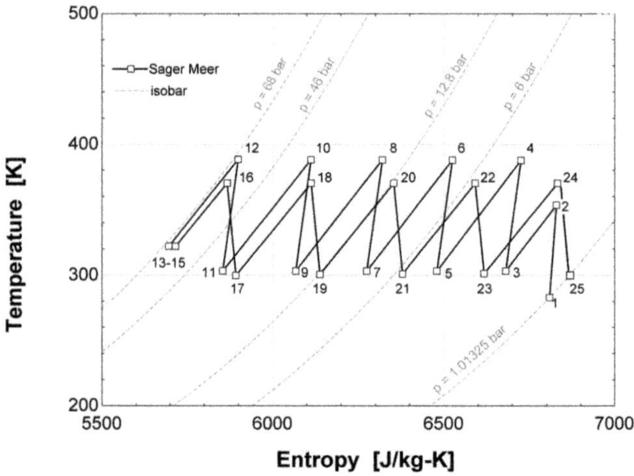

FIGURE 3.22: T-s diagram of quasi-isothermal Sager Meer CAES

Technical work and heat. For the compression a specific technical work of $w_{t,C} = 498\ kJ/kg$ is needed, while in the expansion $w_{t,T} = -346\ kJ/kg$ is generated. Hence, the net work (absolute value of expansion minus compression work) of the system amounts to $-152\ kJ/kg$, thus, the system is an electrical energy consumer. The heat that is used during inter-cooling (compression) is not dissipated but fed into a local heat distribution grid and sums up to

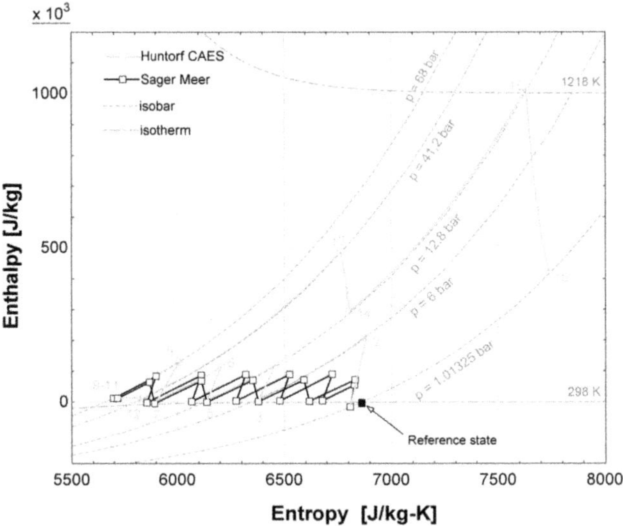

FIGURE 3.23: h-s diagram of quasi-isothermal Sager Meer CAES in comparison with Huntorf (grey)

-471 kJ/kg. The heat taken from the local heat distribution grid for the expansion amounts to 336 kJ/kg.

Power. With the suggested air mass flow rate of 840 kg/s during charge and discharge [Oldhafer et al., 2014; Leithner, 20.01.2015], a compressor electric power of $465\ MW_{el}$ and turbine electric power of $264\ MW_{el}$ is estimated when the mechanical efficiency for both, the compressor and turbine trains, is taken to be 0.9. In [Oldhafer et al., 2014; Leithner, 20.01.2015] better performance values are found since pressure losses and mechanical efficiencies are estimated more optimistically.

Energy storage efficiency. Even though there is no fuel energy added, the Sager Meer concept is not adiabatic, because heat is taken from and delivered to the local heat distribution grid. The assumption is that this heat grid operates at a constant temperature of 370 K hence being able to keep the compressed air stream at this temperature. If a standing alone hot water storage were to be used, this temperature level could not be kept constant during an entire discharging period since the thermal storage would be drained resulting in falling temperatures (comparable with path 6-6′ of ISACOAST, see Fig. 3.17) and, thus, significantly lower efficiencies. Since $Q_{fuel} = 0$, the thermal efficiency η_{th} (Eq.(2.9)) is not applicable and the heat rate values are equal to zero. If the cogeneration is neglected, all efficiency values collapse to the same figure of 56.9 %, see Table 3.19, which is lower than the originally estimated 72.5 % [Oldhafer et al., 2014; Leithner, 20.01.2015]. When considering the heat management, co generation efficiency defined as the ratio of the turbine work fed into the electrical grid ($346\ kJ/kg \cdot 0.9 = 311.4\ kJ/kg$) plus the heat fed into the thermal grid (471 kJ/kg) to the compression work taken from the electrical grid ($498\ kJ/kg/0.9 = 448\ kJ/kg$) plus the heat taken from the thermal grid (336 kJ/kg), is $\eta(cogeneration) = 0.88$.

	η_{caes} %	η_{th} %	η_{rt1} %	η_{rt2} %	η_{rt3} %	η_{rt4} %	hr_1 $\frac{kWh_{fuel}}{kWh_{electric}}$	hr_2
Sager Meer	56.9	n.a.	56.9	56.9	56.9	56.9	0	0

TABLE 3.19: Comparison of the efficiency values of Sager Meer quasi isothermal CAES

3.4 Comparison and Results

Huntorf & McIntosh. When comparing the two existing CAES plants, Huntorf and McIntosh, it is found that their processes are quite similar, as illustrated by Fig. 3.10. The plant layouts, process temperatures and pressures are comparable. Both plants rely on throttling to enable a constant turbine inlet pressure. The cavern pressures are also comparable with a maximum value of 68 bar (Huntorf) and 75 bar (McIntosh) and a minimum value of 46 bar (Huntorf) and 50 bar (McIntosh). In both cases the compression conversion factor η_{cc} (see Section 2.1.6) is around 5 %. Yet, the overall efficiency values of McIntosh are considerably higher than Huntorf's (see Table 3.20) which is due to the exhaust enthalpy recuperation as well as higher inner and mechanical efficiencies. If the exhaust enthalpy recuperation of the McIntosh plant were not taken into account lower efficiencies would result. Such a calculation has been carried out and the results are presented in Table 3.20 as McIntosh*.

	η_{caes} %	η_{th} %	η_{rt1} %	η_{rt2} %	η_{rt3} %	η_{rt4} %	hr_1 $\frac{kWh_{fuel}}{kWh_{electric}}$	hr_2
Huntorf	39.9	9.6	119.1	66.4	39.9	2.9	10.4	1.7
McIntosh	52.3	22.6	136.1	83.0	72.1	4.5	4.4	1.2
McIntosh*	44.2	17.4	136.1	74.4	53.1	3.5	5.7	1.5
ADELE	54.4	n.a.	54.4	54.4	54.4	54.4	0	0
ISACOAST	50.3	22.3	139.7	81.7	68.6	3.4	4.5	1.3
ISACOAST*	53.9	n.a.	95.1	72.8	64.5	5.2	n.a.	0.8
Sager Meer	56.9	n.a.	56.9	56.9	56.9	56.9	0	0

TABLE 3.20: Comparison of different efficiency values; McIntosh* process without exhaust enthalpy recuperator; ISACOAST* process with lower maximum temperatures; n.a. = not applicable

Advanced concepts. Among the advanced concepts considered, the round trip efficiencies of ISACOAST are the highest. These calculated efficiencies can even be increased when taking into account the use of exhaust gas enthalpy in a steam turbine cycle that has not been considered in the calculations above. Nevertheless, this concept is far from realization since temperatures and pressures exceed today's capabilities of gas turbines, well illustrated by the T-s diagram (Fig. 3.17). It is then meaningful to recalculate ISACOAST process with T_7 temperature limited to 1218 K which corresponds to Huntorf's T_{15} combustion temperature (compare Table 3.4 with Table 3.16). The so calculated efficiencies are presented in Table 3.20 as ISACOAST*. It can be observed that the amount of fuel added to the ISACOAST* process is more than halved and the energy output $W_{el,G}$ is reduced by one third if compared to

ISACOAST. Thus, it turns out that ISACOAST* is a net electric energy consumer ($\eta_{rt1} < 1$) hence the definition of thermal efficiency (η_{th}) and heat rate 1 (hr_1) are not applicable as they result in negative numbers. The calculated cogeneration efficiency of the Sager Meer concept is very high, yet since in none of the other concepts a co generation has been taken into account, the upper value of 88 % may be misleading. The round trip efficiency of 56.9 % is a representative figure that is close to the adiabatic CAES ADELE efficiency of 54.4 %.

As pointed out in the introduction, numerous publications which focus on new CAES concepts use thermodynamics of reversible processes and/or ideal gas (Clapeyron) EOS [Grazzini and Milazzo, 2008; Hartmann et al., 2012a; Kim et al., 2012; Pickard, Hansing, and Shen, 2009]. The availability of Huntorf plant operational data listed in Table 3.2 enables (a) to estimate the inner and mechanical efficiencies (see Tables 3.3 and 3.6, respectively) and (b) to assess the effect of various thermodynamic assumptions on the calculated efficiencies. The latter is presented in Table 3.21 showing the calculated Huntorf plant efficiencies using reversible/irreversible thermodynamics and ideal gas/real gas EOS.

	η_{caes} %	η_{th} %	η_{rt1} %	η_{rt2} %	η_{rt3} %	η_{rt4} %	hr_1 $\frac{kWh_{fuel}}{kWh_{electric}}$	hr_2
Huntorf								
real gas, irrev.	39.9	9.6	119.1	66.4	39.8	2.9	10.4	1.7
real gas, rev.	47.8	28.0	173.8	84.7	68.5	4.1	3.6	1.5
ideal gas, irrev.	42.6	13.5	126.6	70.8	47.8	4.1	7.4	1.6
ideal gas, rev.	50.8	32.0	184.0	89.8	79.1	5.8	3.1	1.4

TABLE 3.21: Comparison of the efficiency values of Huntorf (as calculated above, using irreversible thermodynamics and the EOS for real gases in comparison with values obtained for reversible process and/or ideal gas (Clapeyron) EOS

Irreversible vs. reversible thermodynamics. The assumption of reversibility ($\eta_s = 1$) clearly leads to an overestimation of the overall efficiency. When the air is treated as a real gas and reversible thermodynamics is used, the CAES efficiency η_{caes} is overestimated by around 8 percent points (47.8 % against 39.9 %, see Table 3.21) which corresponds to a relative error of almost 20 %. The largest error can be observed in the thermal efficiency values. Under reversibility assumptions a thermal efficiency $\eta_{th}(reversible) = 28.0$ % is found which is 18.4 percent points higher than the value that results from calculations with irreversible thermodynamics, hence a relative error of 191.6 % applies. Again, round trip efficiency η_{rt1} is ignored since it is not applicable to Huntorf process as it has been pointed out previously. The round trip efficiencies $\eta_{rt2/3/4}$ are 18.3, 28.7 and 1.2 percent points too high corresponding to relative errors of 27.6, 72.1 and 63.3 %, respectively. The heat rates calculated with reversibility assumption are also too optimistic. For air as real gas hr_1 shows a relative error of 65 % (3.6 against 10.4). The relative error of the hr_2 is lower with 12 % (1.5 against 1.7).

These absolute and relative errors are in the same order of magnitude when comparing the results of irreversible and reversible thermodynamics using the ideal gas law.

Real vs. ideal gas. As shown in Table 3.21 the CAES efficiency η_{caes} is overestimated by 2.7 percent points (42.6 % against 39.9 %) when ideal gas law is used. Again, the thermal efficiency η_{th} shows the largest deviation with an absolute difference of 4.9 percent points (13.5 % against 9.6 %) corresponding to the relative error of 51 %. The round trip efficiencies

$\eta_{rt2/3/4}$ are also overestimated when the ideal gas assumption is applied. Hence, the efficiency values η_{caes} and η_{rt2} calculated based on figures from ideal gas EOS are overestimated by approximately 7 % compared to those values calculated with real gas assumptions. The relative errors of the heat rates amount to 29 % for hr_1 or 6 % for hr_2, respectively.

Irreversible thermodynamics and real gas EOS versus reversible thermodynamics and ideal gas EOS. When both simplifying assumptions, namely the ideal gas law and the reversible thermodynamics, are used the calculated efficiencies deviate substantially from the plant values. The Huntorf CAES efficiency is then overestimated by 10.9 percent points (50.8 % against 39.9 %) giving a relative error of 27.3 %; the round trip efficiencies $\eta_{rt2/3/4}$ are 23.4, 39.2 and 2.9 percent point too large which corresponds to relative errors of 35.2, 98.7 and 100 %, respectively. The relative errors of the heat rates amount to 70 % for hr_1 or 18 % for hr_2.

Ambient air conditions. In our calculations a constant ambient air temperature of 283 K is used. Yet, the efficiencies vary with ambient air temperature. For example, on a warm summer day of 313 K, the thermal efficiency (η_{th}) of Huntorf CAES is two percent points lower. For a winter day of 263 K ambient temperature, the efficiency values rise between 1 and 1.5 percent points. This effect would even be higher if the inter-cooling temperature of the compression were coupled to the ambient temperature, which is not the case in the calculations at hand, since a constant cooling water temperature of 283 K is assumed. The ambient pressure is set to 1.01325 bar. Changes in the 0.925 to 1.070 bar range that correspond to extreme weather situations, lead to negligible variations (+/- 0.5 percent points) of the process parameters and efficiency.

3.5 Conclusions

This Chapter presents a detailed thermodynamic analysis of the two existing Huntorf and McIntosh Compressed Air Energy Storage (CAES) plants, as well as advanced adiabatic, isobaric and quasi-isothermal CAES concepts under development. The processes are considered at steady-state and as irreversible with air being treated as a real gas. The calculation for the Huntorf plant, of which several complete sets of measured operational data [Krüger, 27.07.2015] serve to develop and validate the calculation methods, are used to test several thermodynamic assumptions concerning both irreversibility and Equation of State (EOS). These methods are then applied to the advanced CAES systems to evaluate all concepts by using a consistent evaluation methodology.

3.5.1 Thermodynamic Assumptions

Realistic inner efficiencies. Usage of irreversible thermodynamics is crucial for an accurate representation of the CAES processes. The assumption of reversibility leads to an underestimate of process temperatures (see Fig. 3.2) and entails an underestimate of technical work for compression and overestimate of technical work for expansion, resulting in a considerable overestimate of efficiency values. Depending on the actual efficiency definition, relative errors in excess of 100 % can occur if reversible thermodynamics is used. Hence, the inner efficiency (η_s) has to be taken into account and - due to its strong impact - should be chosen carefully. It is estimated that for low pressure compressors (up to 10 bar) an inner efficiency of $\eta_s = 0.90$

or 0.91 can be used. For higher pressures figures of 0.80 to 0.85 are appropriate. Turbine inner efficiencies should be in the 0.88 to 0.91 range for a realistic process design.

Air as real gas. For thermodynamic considerations of a CAES, the use of an equation of state (EOS) that treats the (compressed) air as real gas is advocated [Kaiser, Weber, and Krüger, 2018]. Even though the calculation with Clapeyron EOS delivers a good approximation of the process state points during compression and expansion, the storage is poorly represented and the resulting efficiency values tend to be too high. When Clapeyron EOS has been used, in the Huntorf case, the CAES efficiency η_{caes} and round trip efficiency η_{rt2} are overestimated by 2.7 (42.6 % against 39.9 %) and 4.4 (70.8 % against 66.4 %) percent points, respectively, which corresponds in both cases to 7 % relative error.

3.5.2 Efficiency

To characterize energy storage facilities the round trip efficiency is a helpful information. For an adiabatic CAES, where no fuel is added to the process, the efficiency is simply defined as ratio of output to input electrical energy. Such an efficiency is simple and unambiguous. Yet, in fuel-driven CAES concepts, due to the input of both, fuel and electrical energy, there is no unambiguous round trip electrical energy storage efficiency. Thus, an efficiency of fuel-driven CAES is not a self-explanatory figure, but has to be supplemented by the calculation method. Different commonly used efficiency definitions have been examined and their drawbacks have been identified (see Table 2.3).

Fuel-driven CAES. A 'pragmatic' round trip efficiency η_{rt4} was introduced which is the ratio of the electrical energy returned to the grid during repowering to the energy taken from the grid during charging. For Huntorf CAES plant this efficiency is around 3 % while for McIntosh a figure of 4 % is applicable. The reason for such low figures is the thermodynamic inefficiency of compression during which 95 % of the electrical energy taken from the grid is dissipated into heat. Hence, only 5 % is stored in the compressed air which is then converted in the turbine train into electricity. The turbine conversion coefficient (η_{tc}) of Huntorf amounts to 0.59, thus, η_{rt4} round trip efficiency, as defined in Eq.(2.14), is 3 %.

For McIntosh CAES plant the efficiency is a bit larger ($\eta_{rt4} = 4$ %) since the turbine conversion coefficient (η_{tc}) is as high as 0.80, due to the exhaust gas enthalpy recovery as well as a large mechanical efficiency of 0.97.

Adiabatic CAES. It is then obvious that storage of compression heat and its recuperation is necessary to increase the overall energy storage efficiency. Hence, further research in adiabatic CAES (ACAES) is advocated. Yet, the calculations show that efficiency values for an ACAES system, such as the 54 % figure for the ADELE project, are far from the often cited 70 % goals and even further from the the 80 % figure applicable to Pumped Hydro Energy Storage. Even if one assumes a perfect heat storage with a complete heat recovery the ACAES round trip efficiencies are around 66 %.

Isothermal CAES. Another option to overcome the waste of heat during compression is the development of near isothermal compression systems. Thus, quasi-isothermal CAES that limits the maximum temperatures during compression by a large number of compression stages can be applied. The heat removed during inter-cooling is stored in analogy to adiabatic plant schemes. The resulting $57\ \%$ efficiency is in the same order of magnitude as those calculated for the ACAES plant examined. As a matter of fact isothermal compressions have been under development [McBride, Bell, and Kepshire, 2013; Bollinger, 2015] and for a prototype $1.5\ MW$ system an efficiency of 57 % has been quoted [Bollinger, 2015] which is in line with the estimates.

Isobaric CAES. An isobaric CAES concept was examined that avoids throttling of compressed air by using an isobaric air storage reservoir. For the Huntorf plant the effects of such a storage type would be a rise of CAES efficiency by 0.8 percent points (40.7 % against 39.9 %). Thus, it is to question whether such a small efficiency benefit justifies the extra complexity and costs of an isobaric air storage system. Yet, underwater CAES solutions such as developed by Hydrostor, Inc. [VanWalleghem, 2015], Pimm, Garvey, and Drew [2011] and Wang et al. [2016] offer new opportunities for CAES.

Chapter 4

Time Dependent Thermodynamics of CAES

Three aspects make CAES inherently time dependent: (1st) **Isochoric air storage** in underground salt caverns is the state of art for storing compressed air in CAES systems used in both existing CAES plants Huntorf and McIntosh. Thus, the charging process is inherently time dependent since the output pressure of the compressor train (which corresponds to the cavern pressure) is continuously rising. During discharge, on the other hand, cavern pressure is often smoothed via a throttle. Hence, the turbine train is less affected by the cavern pressure as long as the minimum pressure for full load operation is maintained within the cavern. Below this minimum pressure for full load operation, turbines can still be operated in a part load. Such an operation depends on cavern pressure i.e. is also time dependent. (2nd) When CAES is used to balance the **intermittent power supply of renewable energy sources** like wind and solar power flexible operation is required. This entails frequent start-up and run down procedures as well as operation in a part load which will affect the energy efficiencies. Hence, time dependent analysis is required. (3rd) Eventually, for adiabatic CAES systems (ACAES) the temperature of **thermal energy storage** (TES) units is continuously changing during operation which has a considerable effect on the overall plant operation and its efficiencies. Thus, again, dynamic analysis is inevitable.

These three time-dependent aspects and corresponding calculation methods are presented in the following. Calculation methods are validated where possible with measured data of the Huntorf CAES process (Appendix A) or data presented in literature.

4.1 Compressed Air Storage Cavern (CAS)

The charging level of an isochoric compressed air storage cavern (CAS) is in the full load pressure range proportional to the cavern pressure. However, if pressure drops below full load operational pressure of the expander turbines, the correlation of pressure and charging level is no longer linear. This information is further decisive to determine energy capacity and maximum operation duration of the CAES plant. Thus, to determine important key figures like charging level, operation duration, and energy capacity it is imperative to calculate the CAS pressure.

To calculate the air pressure of the cavern the air temperature inside the cavern has to be determined. This can be done by solving the **mass balance** (see Subsection 4.1.1) and the **energy balance** (Subsection 4.1.2) of the air storage system which is shown in Fig. 4.1. The air mass flow rate and heat flow rate are defined to have a positive sign when entering the cavern and negative when leaving it. Appropriate values for the **heat transfer** (heat flow rate $\dot{Q}$ to the surrounding rock) have to be made (Subsection 4.1.3).

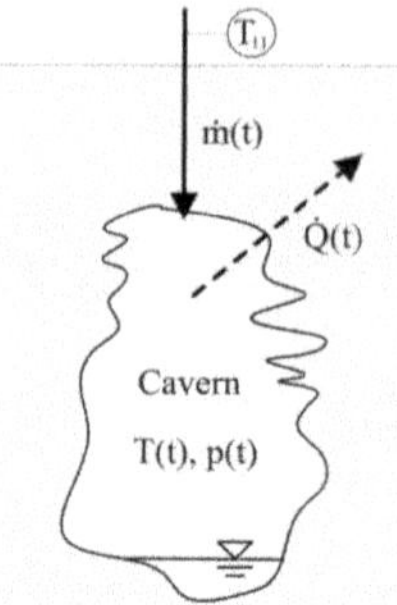

FIGURE 4.1: Compressed air storage cavern (CAS) with air mass flow rate $\dot{m}(t)$ into the storage void and heat flow rate $\dot{Q}(t)$ to the surrounding salt rock

During charging the CAS with the mass flow rate $\dot{m}(t)$ the air inside the storage place heats up due to compression of the stored gas. During discharging the contrary applies due to gas expansion. If the air temperature inside the cavern over time $T(t)$ differs from the surrounding rock temperature T_{rock} heat is transferred according to the general equation $\dot{Q}(t) = k \cdot A \cdot (T(t) - T_{rock})$ (with k being the thermal transmittance; and A the heat transfer surface). It is assumed that the properties of air inside the cavern are homogeneous and can be described by a single bulk property value, as investigated in detail by [Nieland, 2004].

The two theoretically limiting cases are an isothermal cavern ($T(t) = const.$ for perfectly heat-conducting rock) or an adiabatic cavern (perfectly isolated rock without heat conduction; $\dot{Q} = 0$). A sample of measured temperature data in comparison with these two calculated limiting cases is displayed in Fig.4.2.

For **validation** of the calculation methods the results are compared to the measured data presented in Tables A.4 [Krüger, 27.07.2015] and A.5 [Quast and Crotogino, 1979] (see Subsection 4.1.4). Appropriate methods to calculate capacity and the **charging level** (e.g. for energy system simulation) are then presented at the end of this section (in Subsection 4.1.5).

4.1.1 Mass balance

For an air tight storage place the air mass flow rate into or out of the cavern is the only time dependent element. The mass balance is then:

$$\frac{dm}{dt} = V\frac{d\rho}{dt} = \dot{m}(t) \tag{4.1}$$

To calculate the air mass inside the cavern Eq. 4.1 has to be solved for m(t). The two trivial solutions are constant mass flow rate (i.e. $\dot{m}(t) = const.$) or no mass flow rate (i.e. $\dot{m}(t) = 0$) which result in a mass $m(t) = \int \dot{m}(t)dt = \dot{m} \cdot t + m_0$ or $m(t) = m_0$, respectively, where m_0 is the initial air mass inside the cavern which has to be calculated with the starting conditions.

In this thesis either a measured or a constant mass flow rate $\dot{m}$ is used in the simulation. However, the mass flow rate is rather a function of cavern pressure.

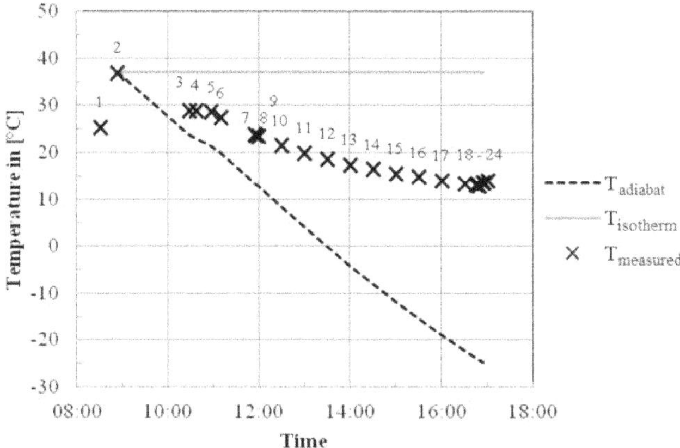

FIGURE 4.2: Measured temperature over time $T_{measured}$ for a discharging test of the Huntorf CAES cavern (Table A.4 [Krüger, 27.07.2015]) in comparison with the two theoretically limiting cases: cavern air temperatures for an isothermal or adiabatic compressed air storage cavern

4.1.2 Energy balance

The compressed air storage cavern is an open system without kinetic or potential energy. Time dependent changes of the inner energy (U) of compressed air inside the storage place can be estimated via the energy balance. It changes with the energy of the mass flow $\dot{m}$ and with the heat flow $\dot{Q}$ crossing the control border. The mass flow may contain enthalpy, kinetic, and potential energy, but in our case only enthalpy is of relevance. Thus, the following equation for time dependent changes of the inner energy is obtained:

$$\frac{dU}{dt} = \dot{m} \cdot h_{\dot{m}} - \dot{Q} \tag{4.2}$$

with $\dot{m}$ = the mass flow rate into or out of the cavern, $h_{\dot{m}}$ = the enthalpy of the air mass flow, and $\dot{Q}$ = heat flow rate to or from the surrounding rock. These values are functions of time (i.e. $\dot{m} = \dot{m}(t)$; $h_{\dot{m}} = h_{\dot{m}}(t)$; and $\dot{Q} = \dot{Q}(t)$). The notation $f(t)$ is omitted for reasons of readability.

The specific inner energy ($u = \frac{U}{m}$) can be substituted into Eq. 4.2:

$$\frac{d(u \cdot m)}{dt} = \dot{m} \cdot h_{\dot{m}} - \dot{Q}$$
$$u\frac{dm}{dt} + m\frac{du}{dt} = \dot{m} \cdot h_{\dot{m}} - \dot{Q}$$

With the terms for mass flow rate ($\frac{dm}{dt} = \dot{m}$):

$$m\frac{du}{dt} = \dot{m} \cdot (h_{\dot{m}} - u) - \dot{Q}$$

...and the terms for specific inner energy ($u = h - p \cdot v$) :

$$m\frac{du}{dt} = \dot{m} \cdot (h_{\dot{m}} - h + p \cdot v) - \dot{Q}$$

$$\frac{du}{dt} = \frac{\dot{m} \cdot (h_{\dot{m}} - h + p \cdot v) - \dot{Q}}{m} \tag{4.3}$$

With the notation of time dependency in Eq. 4.3: $\frac{du}{dt} = \frac{\dot{m}(t)\cdot(h_{\dot{m}}(t)-h(t)+p(t)\cdot v(t))-\dot{Q}(t)}{m(t)}$

This formulation of the energy balance (Eq. 4.3) is valid for charging, discharging, and storage modes. For discharging the enthalpy of the mass flow ($h_{\dot{m}}$) leaving the cavern is equal to the enthalpy of compressed air inside the cavern (h), i.e. $h_{\dot{m}} = h$, thus, in Eq. 4.3 the term $h_{\dot{m}}(t) - h(t)$ collapses to zero in discharging mode. In storage mode the mass flow rate $\dot{m}(t) = 0$ and, thus, the energy balance 4.3 collapses to $\frac{du}{dt} = \frac{-\dot{Q}(t)}{m(t)}$.

Numerical Solution for air as Real Gas

The energy balance (Eq. 4.3) can be solved by iteration and numerical methods with initial values for cavern air temperature T_0 and pressure p_0. Further, the equation for the heat flow rate is $\dot{Q}(T,t) = k \cdot A \cdot (T(t) - T_{rock})$ where rock salt temperature T_{rock}, heat transfer surface A, and thermal transmittance k are known as boundary conditions. The latter can also be calculated as a function of cavern air temperature (such as in [Schwoeppe, Gose, and Scholz, 2008]). An overview of initial and boundary conditions is given in Table 4.3 for the Huntorf case. The numerical method is valid for small time steps $dt = t(i+1) - t(i)$.

The problem has been solved by using MATLAB to handle the large number of unknowns in conjunction with EES (Engineering Equation Solver) to provide properties for air as real gas according to [Lemmon et al., 2000]. Properties of compressed air are calculated for each time step $t(i)$, starting with the initial and boundary conditions. Calculation then advances to the next time step $t(i+1)$ with $\Delta t = 5min$. The formulation of Eq. 4.3 in MATLAB notation is:

$$u(i+1)/(t(i+1) - t(i)) =$$

$$u(i) + (\dot{m}(i) * (h_{\dot{m}}(i) - h(i) + p(i) * v(i)) - k * A * (T(i) - T_{rock}))/m(i) \tag{4.4}$$

In this simulation the mass flow rate $\dot{m}(i)$ is treated as input signal. A first iteration is done with the results from the simplified analytical solution for ideal gas (see following sections) and then repeated to receive a solution for air as real gas.

Numerical Solution for air as Ideal Gas

If one assumes that the compressed air behaves like an ideal gas the energy balance (Eq.4.3) can be further resolved because inner energy and enthalpy are then a function of temperature only. For ideal gas the Clapeyron Equation of State (Eq.4.5) and the Equations 4.6 and 4.7 apply.

$$p \cdot v = R_s \cdot T \tag{4.5}$$

$$du = c_v dT \tag{4.6}$$

$$dh = c_p dT \tag{4.7}$$

The specific heat at constant pressure and at constant volume, c_p and c_v respectively, vary in the temperature range considered from 283 K to 353 K in a small range only. For air as ideal gas $c_p(283K) = 1004J/kgK$ to $c_p(353K) = 1008J/kgK$ and $c_v(283K) = 717J/kgK$ to $c_v(353K) = 721.3J/kgK$. To further simplify, both specific heat capacities are assumed to be constant and summarized in Table 4.1. By integrating dh from a reference temperature

T_{ref} to the compressed air temperature T the correlation $h = c_p \cdot (T - T_{ref})$ is applicable and the following formulation of the energy balance (Eq. 4.3) appears for air as ideal gas with constant specific heat:

$$m \cdot c_v \tfrac{dT}{dt} = \dot{m}[c_p((T_{\dot{m}} - T_{ref}) - (T - T_{ref})) + p \cdot v] - \dot{Q}$$

$$m \cdot c_v \tfrac{dT}{dt} = \dot{m}[c_p(T_{\dot{m}} - T_{ref} - T + T_{ref}) + R_s \cdot T] - k \cdot A(T - T_{rock})$$

$$m \cdot c_v \frac{dT}{dt} = \dot{m}[c_p(T_{\dot{m}} - T) + R_s \cdot T] - k \cdot A(T - T_{rock}) \tag{4.8}$$

with $T_{\dot{m}}$ = the temperature of the air mass flow entering or leaving the cavern. (Again, notation of time dependency is omitted for handier depiction. In more detail one has to write $m(t) \cdot c_v \frac{dT}{dt} = \dot{m}(t)[c_p(T_{\dot{m}}(t) - T(t)) + R_s \cdot T(t)] - k \cdot A(T(t) - T_{rock}))$. This differential equation can be further resolved:

$$m \cdot c_v \tfrac{dT}{dt} = \dot{m} \cdot c_p \cdot T_{\dot{m}} - \dot{m} \cdot c_p \cdot T + \dot{m} \cdot R_s \cdot T - k \cdot A \cdot T + k \cdot A \cdot T_{rock}$$

$$m \cdot c_v \tfrac{dT}{dt} = (\dot{m}(R_s - c_p) - kA) \cdot T + \dot{m} \cdot c_p \cdot T_{\dot{m}} + k \cdot A \cdot T_{rock}$$

$$\tfrac{dT}{dt} = \tfrac{\dot{m}(R_s - c_p) - k \cdot A}{m \cdot c_v} \cdot T + \tfrac{\dot{m} \cdot c_p T_{\dot{m}} + k \cdot A \cdot T_{rock}}{m \cdot c_v}$$

with $R_s - c_p = -c_v$

$$\frac{dT}{dt} = \frac{-c_v \cdot \dot{m} - k \cdot A}{m \cdot c_v} \cdot T + \frac{\dot{m} \cdot c_p \cdot T_{\dot{m}} + k \cdot A \cdot T_{rock}}{m \cdot c_v} \tag{4.9}$$

TABLE 4.1: Assumptions for air as ideal gas (T = 313 K)

Specific heat at constant pressure (c_p)	1005 J/kgK
Specific heat at constant volume (c_v)	718 J/kgK
Gasconstant for air as ideal gas (R_s)	287 J/kgK

Eq.4.9 can be solved numerically with Euler's Method (Explizites Euler Verfahren) or more sophisticated numerical methods (e.g. explicit midpoint method or modified Euler method (Verbessertes Euler Verfahren), and Runge-Kutta methods).

Once the air cavern temperature $T(t)$ is calculated temperature dependent heat capacities c_p and c_v can be calculated as temperature dependent values and used in MATLAB's numerical solution ($c_p(i)$ and $c_v(i)$ instead of constant values) in order to iteratively improve the accuracy of the solution. The iteration has to be repeated several times.

Simplified Analytical Solution for Ideal Gas

The differential equation 4.9 corresponds to the general form $\frac{dT}{dt} = a(t) \cdot T + s(t)$ with $a(t) = \frac{-c_v \cdot \dot{m} - k \cdot A}{m \cdot c_v}$ and $s(t) = \frac{\dot{m} \cdot c_p T_{\dot{m}} + k \cdot A \cdot T_{rock}}{m \cdot c_v}$. It is valid for all three operation modes (charging, discharging with $T_{\dot{m}} = T$, and storage with $\dot{m} = 0$). It can be resolved for T(t) when $a(t)$ and $s(t)$ are constant via $T = T_{homogen} + T_{partikular}$. Thus, an additional assumption has to made: $m(t) = const.$ Further, the solution has then to be divided into charging and discharging modes. For $a(t) = const.$ and $s(t) = const.$ the following general solution is applicable:

$$T_{homogen}(s = 0) = e^{a \cdot t} \text{ (thus } T'_{homogen} = a \cdot e^{a \cdot t}\text{) and } T_{partikular} = s/a$$

$$T = (T_o + \tfrac{s}{a})e^{a \cdot t} - s/a$$

Charging:

$$T(t) = (T_o + \frac{\dot{m} \cdot c_p T_{\dot{m}} + k \cdot A \cdot T_{rock}}{-c_v \cdot \dot{m} - k \cdot A})e^{\frac{-c_v \cdot \dot{m} - k \cdot A}{m \cdot c_v} \cdot t} - \frac{\dot{m} \cdot c_p T_{\dot{m}} + k \cdot A \cdot T_{rock}}{-c_v \cdot \dot{m} - k \cdot A} \tag{4.10}$$

Discharging:

$$T(t) = (T_o + \frac{k \cdot A \cdot T_{rock}}{R_s \cdot \dot{m} - k \cdot A})e^{\frac{R_s \cdot \dot{m} - k \cdot A}{m \cdot c_v} \cdot t} - \frac{k \cdot A \cdot T_{rock}}{R_s \cdot \dot{m} - k \cdot A} \tag{4.11}$$

This analytical solution based on the afore mentioned simplifications (ideal gas, constant heat capacities, constant air mass inside the cavern) was proposed by [Raju and Khaitan, 2012; Xia et al., 2015]. However, Xia et al., 2015 use a set of three different sets of solutions to distinguish charging, discharging, and storage. The latter is not required since the relevant terms collapse to zero for $\dot{m} = 0$.

4.1.3 Heat Transfer - Thermal Properties of Salt Rock

Schön [2015] compiled different measured values of the heat conductivity (λ) of salt rock at different sites. He [Schön, 2015] finds values in the 1.55 to 5.34 W/Km range and uses as mean value 5.8 W/Km. Kushnir, Ullmann, and Dayan [2012a] use (with reference to [Schön, 2015]) a mean value of $\lambda = 4\ W/Km$ and a specific heat $c = 840\ J/kgK$, whereas according to [Leuger and Beutel, 2012] the heat conductivity (λ) of salt is in the range 5.0 to $5.5\ W/Km$ and heat capacity (c) is in the $850 - 890\ J/kgK$ range. Since thermal properties of rock salt are site dependent the rounded values in Table 4.2 are used as estimates.

TABLE 4.2: Assumptions for thermal properties of salt rock

Heat conductivity (λ)	5 W/Km
Heat capacity (c)	850 J/kgK

For the term $\dot{Q}(T) = k \cdot A(T - T_{rock})$ reasonable assumptions for the heat transfer with the surrounding rock have to be made. Since the thermal conductivity in salt rock is around $5\ W/Km$ (see Table 4.2) the thermal transmittance (whose inverse is calculated as sum of the inverse heat transfer coefficient α and heat conductivity λ as $\frac{1}{k} = \frac{1}{\alpha} + \frac{L}{\lambda}$) is expected to be in the same order of magnitude or lower. The cavern wall surface (A) acts as heat transfer surface and is known to be larger than a cylinder with the same volume [Crotogino, Mohmeyer, and Scharf, 2001-04-15]. A surface factor $z > 1$ is introduced, see Fig. 4.3. Hence, the surface (A) can be expressed as $A = z \cdot \pi \cdot r^2 \cdot h$, with z to be determined by fitting of measured and calculated data.

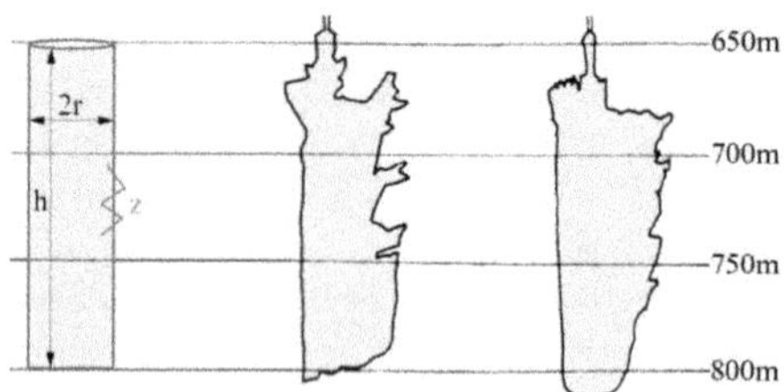

FIGURE 4.3: Cavern shape assumptions (left) and actual Huntorf air storage caverns (adapted from [Quast and Crotogino, 1979])

The rock salt temperature is set to $T_{rock} = 313\ K (= 40°C)$ according to [Quast and Crotogino, 1979].

4.1.4 Validation of CAS Calculation Methods with Huntorf Data

The calculation methods presented above are compared to the two sets of measured data from the Huntorf CAES plant for validation: a data set measured in 2011 representing a discharging trial where both caverns are in use, see Table A.4 [Krüger, 27.07.2015]; and data from commissioning trials in 1978 with one of the two caverns only, see Table A.5 [Quast and Crotogino, 1979]. Results are presented in Fig. 4.4 and 4.5, respectively. The temperatures are measured at the cavern head above ground. Therefore accuracy of this comparison is limited. The initial and boundary conditions as well as constants and geometry of the caverns used for these solutions are presented in Table 4.3.

TABLE 4.3: Initial and boundary values, constants, and geometry of the compressed air storage caverns in Huntorf CAES for the data from Quast and Crotogino [1979] and Krüger [27.07.2015] as well as the Status Quo

Data	Quast	Krüger	Status Quo
initial cavern temperature (T_0)	318 K	316 K	
rock salt temperature (T_{rock})	313 K	313 K	313 K
thermal transmittance (k)	3 W/Km	3 W/Km	3 W/Km
surface factor (z)	10	10	10
cavern height (h)	150 m	150 m	150 m
number of caverns	1	2	2
cavern volume (V)	135,000 m^3	254,000 m^3	252,000 m^3
heat transfer surface (A)	160,000 m^2	309,000 m^2	309,000 m^2

The cavern temperatures and pressures for the two limiting cases isotherm and adiabatic storage cavern are illustrated in Fig. 4.4 and 4.5 as well. It appears that the assumption of an adiabatic cavern is far from the actually measured data. For the discharging trial (see Fig. 4.5, [Krüger, 27.07.2015], Table A.4) the adiabatic cavern air temperature drops below $0°C$. Thus, the resulting pressures are not representative.

For isothermal storage cavern, on the other hand, the resulting pressures are accurate within a 2 bar error range. Thus, for general estimates the isothermal simplification appears as valid option with a certain error range, while adiabatic cavern assumption is rather not.

The thermal transmittance k is set to 3 W/Km^2. The total cavern surface in the Huntorf example is estimated to be $A = 309,000\ m^2$. The total cavern volume is set to $V = 254,000\ m^3$. This value stands in contrast to the often cited literature value of $310,000\ m^3$ cavern volume for Huntorf. However, these calculation results have lately been confirmed by measuring the cavern volume via sonar surveying. It appears that the simplified analytical solution despite its extensive assumptions (such as constant air mass inside the cavern) results in comparatively accurate prediction of both cavern air temperatures and pressures. Since measured air temperatures are only accurate within an estimated $+/-5\ K$ range and since air pressures are the more important output figure for capacity and charging level estimates, the computational complexity of a numerical solution for real gas seems to be unjustified. To calculate cavern air temperatures and pressures it is to advice to use the lean solutions for ideal gas (numerical or analytical).

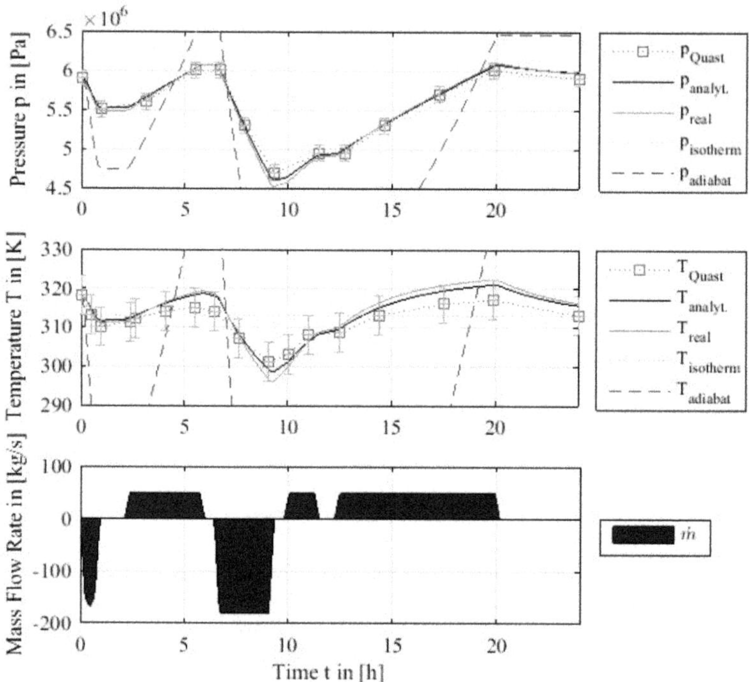

FIGURE 4.4: Cavern pressure, temperature, and mass flow rate over time for measured data (from Quast and Crotogino, 1979) of the Huntorf CAES plant compressed air storage cavern

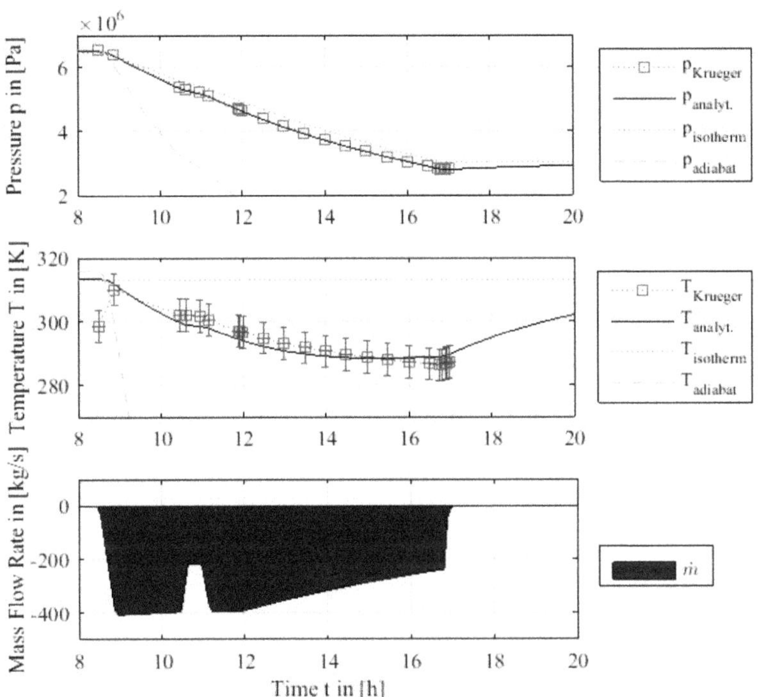

FIGURE 4.5: Cavern pressure, temperature, and mass flow rate over time for measured data (from Krüger, 27.07.2015) of the Huntorf CAES plant compressed air storage cavern

4.1.5 Charging Level of the CAS for Energy System Analysis

In energy systems analysis usually residual load data or data on market signals generate a virtual operation schedule for the energy storage facilities considered. One limiting factor for such an operation schedule is the charging level of the storage plant, i.e. for CAES the charging level of the compressed air storage cavern (CAS). When considering only the full load operation range, the charging level is approximately proportional to the cavern pressure: the minimum operation pressure corresponds to 0% and the maximum operation pressure corresponds to 100%. For Huntorf these values are 46 bar (0%) and 72 bar (100%) cavern pressure. However, Huntorf CAES plant can also operate at pressures below 46 bar. It then operates in a part load, i.e. the gas turbines operate with a lower inlet pressure and different air mass flow rates and temperatures. The power output is then lower and it appears that the charging level is no longer proportional to the cavern pressure. Instead the work W that can be generated with the actual cavern pressure has to be determined to identify the charging level. To do so a virtual complete discharging in the power range 72 bar to 30 bar is simulated with the validated calculation methods presented above. Further appropriate assumptions for the power output are taken from the measured data. The power generated by the turbines P is a function of the air mass flow rate $\dot{m}$ and the technical work w_t and the mechanical efficiency η_{mech}, see Eq. 4.12.

$$P = \dot{m} \cdot \eta_{mech} \cdot w_t \tag{4.12}$$

The technical work w_t (defined according to Eq. 4.13) and the mechanical efficiency have been explored in detail in Chapter 3 as well as [Kaiser, Weber, and Krüger, 2018].

$$\delta w_t = v \cdot dp \tag{4.13}$$

To determine the work of the overall process power has to be integrated over time (Eq. 4.14).

$$W = \int_{t_0}^{t_{end}} P \, dt \tag{4.14}$$

The results are presented in the Figures 4.6, 4.7 and 4.8.

The complete discharging trial depicted in Fig.4.6 illustrates how the thermodynamic cavern air temperature behavior adds on some extra complexity to determining the storage capacity. During discharging the CAS air temperatures drop by around 30 K. Once minimum pressure is reached (here 30 bar) turbine operation stops. Via heat transfer from the cavern walls, air temperature starts to rise which also increases cavern air pressure. After a few hours (here approximately 9 hours) the cavern air pressure is again sufficiently high for a short start-up of the turbine train adding some extra capacity to the system. This extra capacity is around 2 GWh or 7 % of the total capacity. However, this extra capacity cannot be retrieved at once, but requires some extra storage time to allow for natural convection reheating the cavern air temperature (in the following termed "regeneration time"). This overall issue leads to the fact that the capacity is no unique function of the cavern air pressure but only an assignment, illustrated by Fig.4.8. Depending on when the regeneration time occurs the kink in the assignment depicted in Fig.4.8 may appear at other pressures.

The approximate solution with an isothermal CAS (dotted lines in Fig.4.6 and 4.7) is not able to predict this behavior, but only reflects the total capacity despite the fact that it is not retrievable at once.

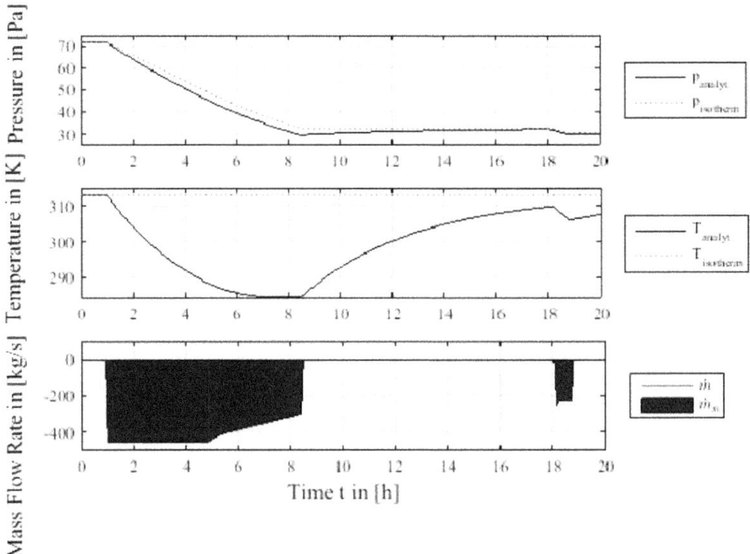

FIGURE 4.6: Cavern pressure, temperature and mass flow rate over time for a simulated complete discharging cycle for the Huntorf CAES configuration

A complete discharging situation in which such unavailable capacity may occur might not be a typical operation mode in the context of intermittent renewable energy generation. Thus, for energy system analysis it is to advice to check if this type of complete discharging occurs in order to handle it properly of simplify the CAES (or more precisely the CAS) model accordingly.

4.2 Compressor (C) and Turbine (T) in Unsteady Conditions

Turbo compressors and turbines are preferably designed for one specific set of operation parameters (mass flow, temperature, pressure). Yet, due to the shifting cavern pressure and the demand for flexible operation when combined with renewables, non-steady operation during start up or run down procedures is an important issue to estimate CAES process characteristics.

4.2.1 Changing Cavern Pressure

Compression. The power consumption of compressors depends on cavern pressure and air mass flow rate of charging. Zhang and Cai [2002] present calculation methods to estimate characteristic performance maps of compressors. The process parameters are used in a dimensionless reduced form. The dimensionless reduced mass flow rate ($\dot{G}$) and the dimensionless reduced rotational speed ($\dot{n}$) are defined as:

$$\dot{G}_c = \frac{\dot{m}_c \cdot \sqrt{T_{cin}}/p_{cin}}{\dot{m}_{c0} \cdot \sqrt{T_{cin0}}/p_{cin0}} \tag{4.15}$$

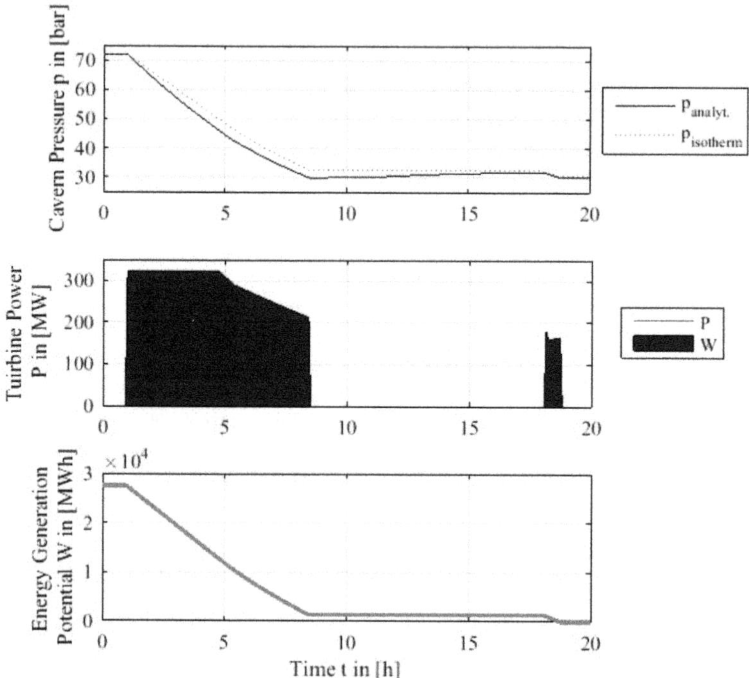

FIGURE 4.7: Cavern pressure, power, and energy over time for a simulated complete discharging cycle

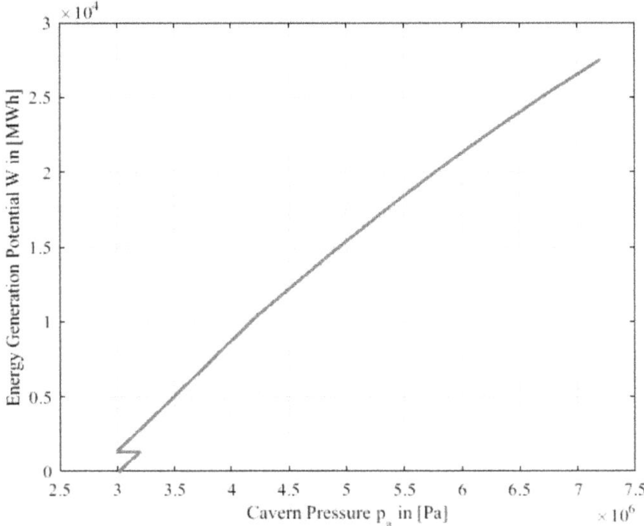

FIGURE 4.8: Charging level (energy) and cavern pressure correlation

$$\dot{n}_c = \frac{n_c/\sqrt{T_{cin}}}{n_{c0}/\sqrt{T_{cin0}}} \tag{4.16}$$

The characteristic performance map of a compressor can then be calculated with Eq.4.17 and Eq.4.18 [Zhang and Cai, 2002; Ebert, 1992] *(p.11)*. Using the parameters $C_x = 0.36$, $C_y = 1.06$ and $C_z = 0.3$ [Zhang and Cai, 2002] and process characteristics of the Huntorf CAES process ($\dot{m}_{c0} = 108\ kg/s$ and $p_{c,0} = 70\ bar$) a performance map is created, see Fig.4.9. However, the actual performance depends on machinery type, air temperature, cooling water temperature, and other factors. Thus, these figures are used as a general estimate to show the interdependencies in unsteady process conditions and are not fully applicable to the existing CAES plants.

$$\dot{\beta} = c_1 \cdot \dot{G}_c^2 + c_2 \cdot \dot{G}_c + c_3 \tag{4.17}$$

$$\dot{\eta}_{s,c} = (1 - C_z \cdot (1 - \dot{n}_c)^2) \cdot (\dot{n}_c/\dot{G}_c) \cdot (2 - \dot{n}_c/\dot{G}_c) \tag{4.18}$$

with $c_1 = \dot{n}/(C_x \cdot (1 - C_y/\dot{n}) + \dot{n} \cdot (\dot{n} - C_y)^2)$; $c_2 = \frac{C_x - 2 \cdot C_y \cdot \dot{n}^2}{C_x \cdot (1 - C_y/\dot{n}) + \dot{n} \cdot (\dot{n} - C_y)^2}$; and $c_3 = -\frac{C_x \cdot C_y \cdot \dot{n} - C_y^2 \cdot \dot{n}^3}{C_x \cdot (1 - C_y/\dot{n}) + \dot{n} \cdot (\dot{n} - C_y)^2}$.

Expansion. The power output of the turbines on the other hand is widely unaffected by changing cavern pressure since the input pressure of the turbines is kept constant through a throttle. Only if the cavern pressure drops below outlet throttle pressure, the output power of the turbines falls. To describe the turbine performance a comparable approach as for compressors can be established with Eq.4.19 and 4.20 with $C_z = 0.3$ [Zhang and Cai, 2002].

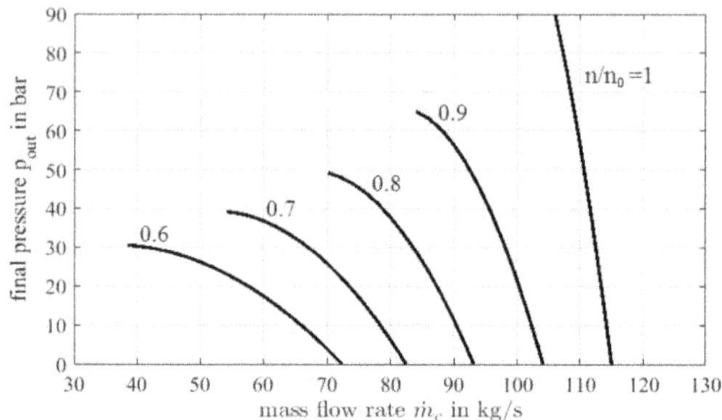

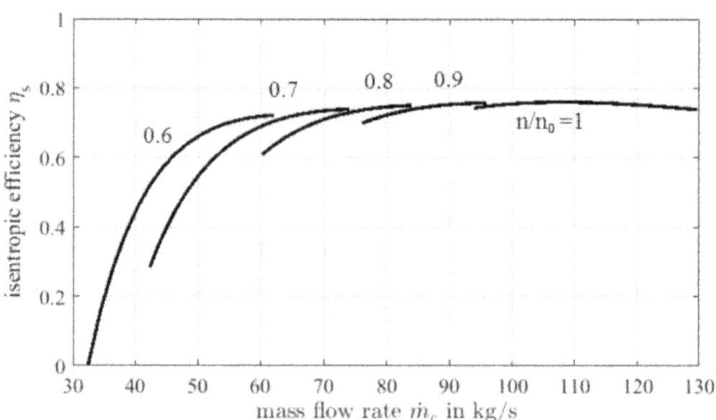

FIGURE 4.9: Performance map of a compressor

$$\dot{\beta} = (\dot{G}_t/\alpha)^3 \cdot T_{tin}/T_{tin0} \cdot p_{it0}^2 - (\dot{G}_t/\alpha)^2 \cdot T_{tin}/T_{tin0} + 1)/(p_{it0}^2) \tag{4.19}$$

$$\dot{\eta}_{s,t} = (1 - C_z \cdot (1 - \dot{n}_t)^2) \cdot (\dot{n}_t/\dot{G}_t) \cdot (2 - (\dot{n}_t/\dot{G}_t)) \tag{4.20}$$

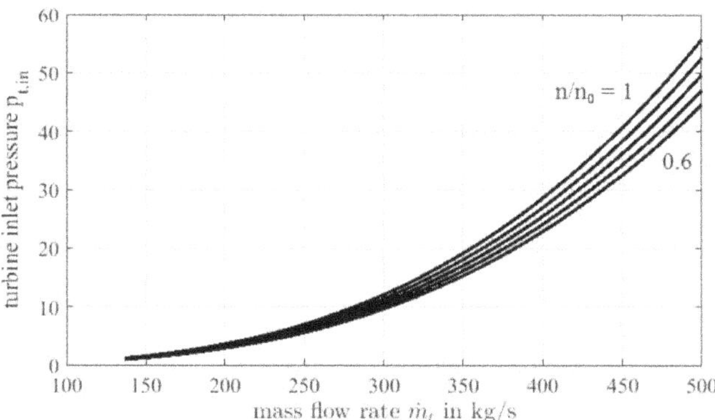

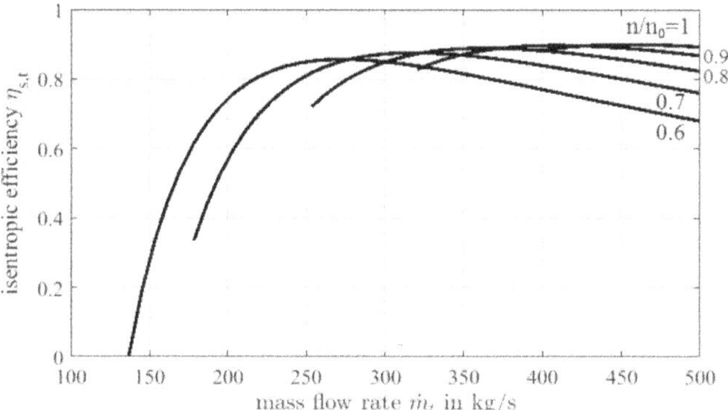

FIGURE 4.10: Performance map of a turbine

Mass flow rate $\dot{m}_t = 455\ kg/s$ of the Huntorf plant is used as parameter. However, these figures only show a general correlation. The actual plant design and optimization of such a system can lead to other characteristic maps.

4.2.2 Start-Up/Run Down Procedures

Start-up and run down are usually undesirable sub-optimal states for turbo machinery, because technical work, efficiency and exit temperatures tend to different optima. Information on start-up of charging and discharging mode as well as on part load behavior of Huntorf CAES is presented by Hoffeins, Romeyke, and Sütterlin [1980]. Start-up and switch time duration are given in Table 4.4. When considering an operation schedule with long steady state operation patterns these effects can be neglected. However, in a wind and solar power dominated market very short operation duration often apply. In such a scenario, start up and run down procedures have a considerable proportion of the operation time and, thus, on efficiency and cost. Thus, these start-up and switch times have to be respected in a time-dependent analyses of the overall system behavior.

TABLE 4.4: Start-up and switch times of the Huntorf CAES plant during commissioning 1978 [Hoffeins, Romeyke, and Sütterlin, 1980]

state change	duration in [min]
gas turbine start-up time (normal)	11
gas turbine start-up time (fast)	6
compressor start-up time	4.5
switch from gas turbine to compression	36
switch from compression to gas turbine (normal)	21
switch from compression to gas turbine (fast)	16

4.2.3 Operation in a Part Load

Compressor. Part load operation of compressors is often possible in a range of (60 or) 70-100 % of the rated power [Lechner and Seume, 2010]. However, this comes along with lower efficiency values. For the considerations at hand no part load operation is considered except the afore-mentioned cavern pressure and start-up procedure related issues. Such a setting corresponds to the Huntorf set up. However, generally speaking, it is possible to use several compressor trains in a parallel arrangement to allow for a more flexible market activity.

Turbine. The power output of the turbines can be reduced by lowering the air mass flow rate or the combustion temperatures and, thus, allows for a relatively flexible market activity. In such modes, less fuel and air is used in the process. However, due to lower inner efficiencies the specific heat and air rates increase with decreasing nominal power. This relation has been calculated for the Huntorf CAES plant based on measured data (see Tables 3.2, A.2, and A.3) and is depicted plant in Fig.4.11.

Specific heat rate (hr_2) (Eq. 2.10, [Kaiser, Weber, and Krüger, 2018]) and specific air rate (ar) are defined as:

$$hr_2 = Q_{fuel}/W_{el,Turb} \tag{4.21}$$

$$ar = W_{el,Comp}/W_{el,Turb} \tag{4.22}$$

With the definitions of hr_2 (Eq.4.23) and ar (4.22) the Huntorf values can be calculated from measured data, see Fig.4.11. To estimate hr_2 and ar based on the nominal turbine power (P in MW) the following approximations can be used (see Fig.4.11, hr_2 and ar)

$$hr_2 = 36,67/P[MW] + 1.4103 \tag{4.23}$$

$$ar = 59.17/P[MW] + 0.786 \tag{4.24}$$

With these characteristic value maps for hr_2 and ar, fuel and air consumption of the turbines can be estimated in the entire part load operation range (100 to 320 MW).

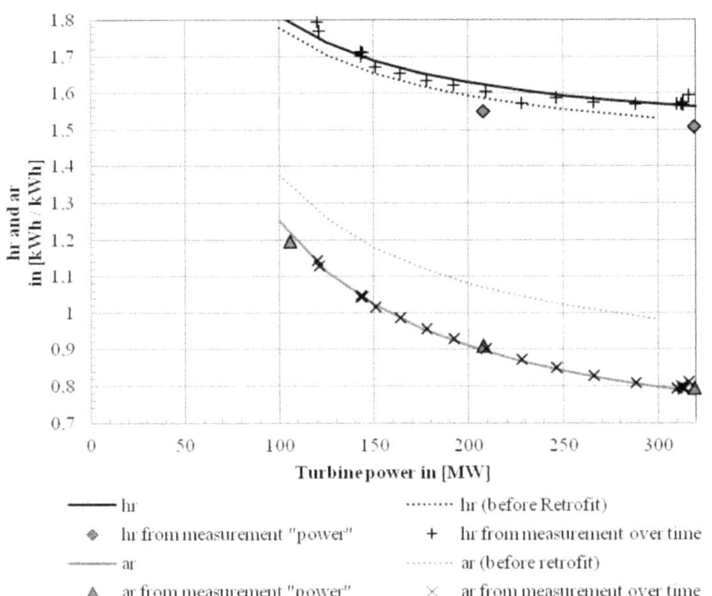

FIGURE 4.11: Specific heat rate (hr_2) and specific air rate (ar) of the Huntorf CAES plant from plant documentation (before retrofit) and based on measured data after the retrofit

4.3 Thermal Energy Storage (TES)

Two phase model packed bed TES. In this section a calculation methods to estimate the thermodynamics of a thermal energy storage (TES) unit is presented. In analogy to [Park et al., 2014] and [Sciacovelli et al., 2017] a packed bed thermal energy storage system is used to estimate transient aspects of this process.

The calculation method, also known as "Schuman Model" [Singh, Saini, and Saini, 2009] is based on a two phase model (air and packed bed). The heat transfer is estimated via Number of Transfer Units (NTU) method as $NTU = h_v AH/(\dot{m} \cdot c_{p,air})$. A finite difference solution is used to solve for the air temperature (T_a, Eq.4.25) and the packed bed temperature (T_b, Eq.4.26). T_a and T_b span over the space vector $\vec{x}$ over the total length h and time vector $\vec{\tau}$. The problem is solved numerically with T_a and T_b as matrices that are solved simultaneously for every place i and every time step j. As initial and boundary conditions the inlet air temperature can be constant or a function of time. The initial temperature of the packed bed can be a constant value or any temperature distribution at will, e.g. the final state of a previous run.

$$T_{a(i,j)} = T_{a(i-1,j)} - (T_{a(i-1,j)} - T_{b(i-1,j)}) \cdot (1 - e^{-\frac{NTU \cdot \Delta x}{H}}) \tag{4.25}$$

$$T_{b(i,j+1)} = T_{b(i,j)} + \Delta\tau \cdot [(1 - e^{-\frac{NTU \cdot \Delta x}{H}}) \cdot \frac{H}{\Delta x} \cdot (T_{a(i,j)} - T_{b(i,j)}) + \frac{k_u \cdot \Delta A}{\dot{m} \cdot c_{p,a}} \cdot (T_{rock} - T_{b(i,j)})] \tag{4.26}$$

The overall thermal transmittance k_u is estimated with the procedures described by Park et al. [2014]. Properties of air are estimated as a function of temperature [Park et al., 2014]. A detailed description of these methods can be found in [Singh, Saini, and Saini, 2009; Park et al., 2014; Sciacovelli et al., 2017].

Validation of TES calculations methods. Several sets of measured data from experimental investigations have been published by Meier, Winkler, and Wuillemin [1991], Anderson et al. [2015], and Cascetta et al. [2015]. The comparison of measured and calculated data shows that the calculation method in use is able to estimate the TES behavior, see Fig.4.12, 4.13, and 4.14.

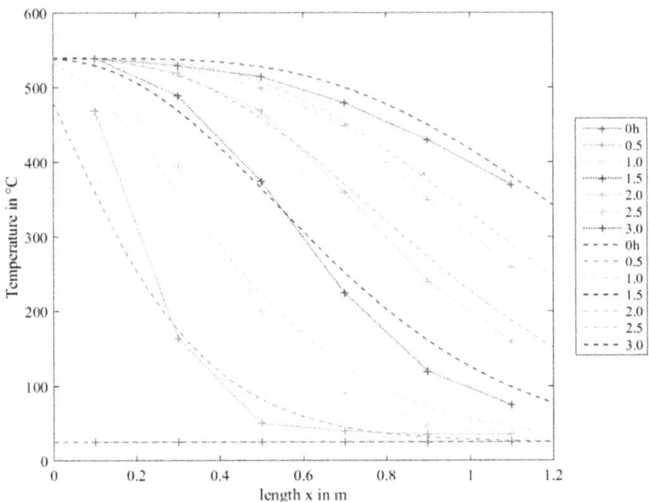

FIGURE 4.12: Validation of TES model with measured data from Meier, Winkler, and Wuillemin [1991] and a thermal transmittance $h_v = 1400 W/m^2K$

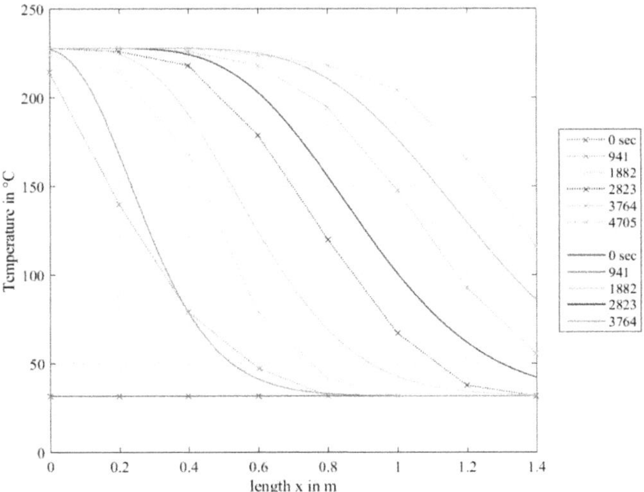

FIGURE 4.13: Validation of TES model with measured data from Cascetta et al. [2015] and a thermal transmittance $h_v = 12000 W/m^2K$

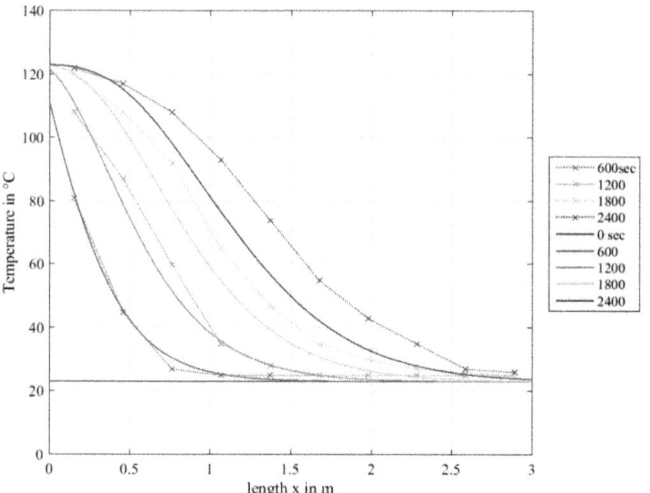

FIGURE 4.14: Validation of TES model with measured data from Anderson et al. [2015] and a thermal transmittance $h_v = 20000 W/m^2K$

Chapter 5

Hydrogen Options for CAES, Huntorf Case Study

Hydrogen (H_2) is a combustible gas that can *inter alia* be generated from water and electrical energy by water electrolysis. In a subsequent oxidation (via combustion or in a fuel cell) water (or steam) is then ideally the only reaction product, see Fig. 5.1. If the energy used for the production of hydrogen is renewable, it can be considered a carbon-free energy cycle without emission of green house gases or use of limited fossil resources. Hence, hydrogen can be a sustainable alternative to fossil fuels. However, the process is not yet fully developed since flame temperature, flame velocity and diffusion coefficient of hydrogen are high and flammable limits are wide and, thus, harder to handle in combustion processes. During combustion with air, NO_x emissions can occur and sophisticated cooling methods have to be applied. An overview of hydrogen properties in comparison with methane (CH_4) and natural gas (NG) as used in the Huntorf CAES process [EWE-Netz GmbH, 2015] is given in Table 5.1. Despite these challenges, several pilot projects have been successfully implemented, in detail investigated by Lund et al. [2015].

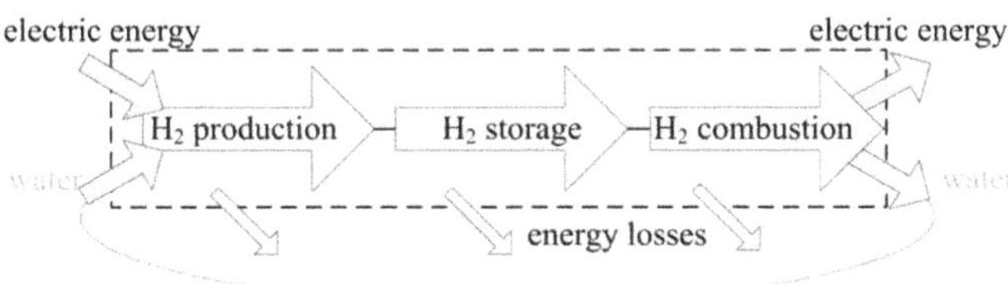

FIGURE 5.1: Main components of a hydrogen energy cycle

5.1 Hydrogen Production

Currently, hydrogen is mainly produced from fossil hydro carbons via steam methane reforming [Turner, 2004]. It is also produced as byproduct in various chemical processes such as production of chlorine, cyanide, ethane and acetylene [Kaiser, 2012-06-28]. But to consider it as a "clean" fuel it has to be generated sustainably from renewable sources such as wind and solar power or biomass. Possible production pathways are electrolysis, thermo-chemical cycles or biomass processing ranging from reforming to fermentation technologies [Turner, 2004]. Furthermore, photo-biological and photo-electrochemical processes are considered in

TABLE 5.1: Properties of hydrogen (H_2) [Leachman et al., 2009], methane (CH_4) [Lee et al., 2010; Sørensen, 2007; Setzmann and Wagner, 1991] and a natural gas (NG) used in Huntorf CAES [EWE-Netz GmbH, 2015]

Property	Unit	H_2	CH_4	NG
Molar mass M	g/mol	2.016	16.03	17.73
Density* ρ	kg/m^3	0.082	0.657	0.793
LCV*	MJ/kg	120	50	40.6
max. burning velocity	cm/s	$289, 346$	$37, 43$	
flammable limits in air		4-75 %	5-15.4 %	
auto ignition	^oC	$500, 585$	537	
diffusion coefficient in air	$10^{-4}m^2/s$	0.61	0.16	

*at T=298.15 K and p=1.01 bar;

recent research activities [Turner, 2004]. These pathways split up into several process variants. A general overview is given in Table 5.2. Detailed investigations on different hydrogen production pathways can be found in [Stolten, 2010].

TABLE 5.2: Production pathways of hydrogen

Process type	process variants (examples)
steam reforming	- methane - coal gasification
byproduct in chemical processes	- production of chlorine, cyanide, ethane and acetylene gases
water electrolysis	- alkaline electrolysis - polymer electrolyte membrane (PEM) - solid oxide electrolysis (SOEC) at high temperatures
thermo-chemical biomass processing	- reforming - thermophilic fermentation - photo-fermentation
photo-biological photo-electrochemical	

5.2 Hydrogen Storage in Salt Caverns

The storage of hydrogen in contrast to natural gas poses several problems due to the low density and high diffusivity of gaseous hydrogen. To achieve high energy densities liquid hydrogen storage as well as ab- and adsorption storage options have been developed e.g. for mobility applications. To store hydrogen in a power plant scale low cost options are desirable such as the storage in underground salt caverns. In Huntorf, natural gas is currently stored in an underground salt cavern by EWE Netze. It is to discuss if this is an option for hydrogen. The thermodynamic properties of such a system that consists of compression including cooling, storage and throttling (Fig.5.2) are to be considered.

Geological aspects of hydrogen storage in salt caverns. The general feasibility of hydrogen storage in salt caverns is proven in operation in Teesside (UK) and in Texas (US) [Crotogino et al., 2010]. At the Huntorf site there are 7 storage caverns in operation, two air storage caverns and 5 natural gas storage caverns [LBEG, 2012]. The salt dome covers approximately

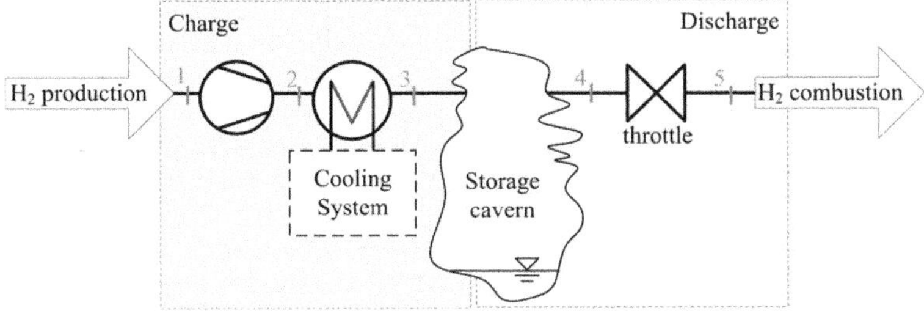

FIGURE 5.2: Hydrogen storage process including compressor, heat exchanger, storage cavern, and throttle

a surface of 1600 ha (16 km^2) at a depth of 1000 m under mean sea level. Its top layer is less than 500 m below surface and extends into depth of up to 6000 m below surface (under mean sea level) [*Geoviewer Bundesanstalt für Geowissenschaften und Rohstoffe, Arbeitsbereich Geodatenmanagement, InSpEE Webdienst Salzstrukturen*]. The salt deposit type is characterized as Zechstein deposit ("kompressiv überpraegtes Diapir"), see Fig. 5.3 [Pollok et al., 2015]. The Huntorf salt dome is classified as suitable for H_2 storage in an investigation of KBB UT, a salt cavern manufacturer company [Klumpp, 2016].

FIGURE 5.3: Scheme of the salt deposit in the Huntorf area

The ceilings of the Huntorf air caverns are located about 650 m below surface and extend to a depth of 800 m [Quast and Crotogino, 1979; Crotogino, Mohmeyer, and Scharf, 2001-04-15]. Permissible pressure ranges are 1 bar to 75 bar and operating pressure is in the 20 to 75 bar range [Quast and Crotogino, 1979]. The natural gas storage caverns extend into depth of up to 1400 m below surface [LBEG, 2012]. In [Crotogino and Hamelmann, 2007] an approximation to estimate allowable pressures (p_{min} and p_{max}) within a cavern based on the depth of the cavern is given: $p_{max}[bar] = 0.18 \cdot depth[m]$ and $p_{min} = p_{max}/3$ In conclusion, the existing salt deposit seems not to be a limiting factor for hydrogen storage underground in the Huntorf area. However, detailed analysis of deposit tightness and other factors are required to authorize such a storage.

Thermodynamics of hydrogen storage. To consider the storage of hydrogen the thermodynamics of a simplified hydrogen storage cycle (see Fig. 5.2) is analyzed using similar methods as used for analyzing the overall compressed air energy storage systems. The initial state $i = 1$ of hydrogen depends on the production method. A commercially available process variant of alkaline electrolysis is chosen that works at high pressures of 30 bar [Stolten, 2010], p. 264. Thus, the initial state of hydrogen is fixed at $p_1 = 30\ bar$ and $T_1 = 298\ K$. An optimization of hydrogen production pathway in combination with compression is beyond the scope of this work. The compression ratio is chosen to be 2, thus, $p_2 = 60\ bar$. The temperature T_2

is estimated via isentropic efficiency $\eta_s = 0.80$ defined as $\eta_s = (h_{2s} - h_1)/(h_2 - h_1)$ where h_{2s} is the enthalpy of state point 2 for a reversible compression. Thus, the temperature of the irreversible compression process can be calculated as function of pressure p_2 and enthalpy h_2 and amounts to $T_2 = 371K$. The allowable storage temperature is 323 K in analogy to the air storage temperature. Hence, the heat exchanger cools the compressed hydrogen to this temperature. The hydrogen storage cavern is assumed to be isothermal so that $T_3 = T_4 = 323\ K$.

In discharging mode, hydrogen is throttled to the combustion chamber inlet pressure of $p_5 = 42\ bar$, which is considered as an isenthalpic process and, thus, comes along with a temperature change according to the Joule-Thompson-Effect which in this case is a slight increase to $T_5 = 324\ K$. For each pressure and temperature specific enthalpy and specific entropy are estimated with appropriate formulas for hydrogen as real gas according to Leachman et al. [2009], see Table 5.3.

TABLE 5.3: State variables of the hydrogen storage cycle

i	pressure p_i in bar	temperature T_i in K	specific enthalpy h_i in kJ/kg	specific entropy s_i in kJ/kg-K
1	30	298.0	13	39.4
2	60	380.2	1220	40.0
3	59	323.0	381	37.7
4	59	323.0	381	37.7
5	42	323.6	381	39.2

A similar thermodynamic calculation for air results in comparable process temperatures. This is due to the similarity of nitrogen (as main component of air) and hydrogen – both are diatomic molecules. Yet, the technical work required for compressing hydrogen is higher. The technical work required for this hydrogen compression amounts to $w_t = h_2 - h_1 = 1207kJ/kg_{H2}$. The heat removed in the subsequent cooling stage $q = h_3 - h_2 = -839\ kJ/kg_{H2}$. To compress the air under the same conditions a technical work of only $w_t = 82\ kJ/kg_{air}$ is required and a heat of $q = -61\ kJ/kg_{air}$ has to be removed. This considerable difference is so due to the low molecular weight (M) of hydrogen. The ratio of molar masses of air to hydrogen amounts to M(air) / M(H_2) = 28.97 / 2.016= 14.4, see Table 5.4. However, the differences cannot only be explained by the molar masses of both gases as the compression conversion coefficient illustrates.

TABLE 5.4: Technical work (w_t) and heat removed (q) in the hydrogen compression using the state variables presented in Table 5.3 compared to the same compression with air

	H_2	Air	ratio Air/H_2
M	2.016 g/mol	28.97 g/mol	14.4
w_t	1207 kJ/kg	82 kJ/kg	14.7
q	-839 kJ/kg	-61 kJ/kg	13.8
η_{cc}	31%	26%	

Compression conversion coefficient. In the thermodynamic analysis of different CAES processes the compression conversion coefficient η_{cc} is defined as ratio of energy contained in the stored gas to electric energy required for compressing it i.e. $\eta_{cc} = (W_{el,M}\eta_{mech} -$

$Q_{loss})/W_{el,M}$. For steady state conditions the specific work and specific heat can be used $\eta_{cc} = (w_{t,C} - q_{loss})/w_{t,C}$. This value can be used to estimate the energy content stored in the compressed gas and is a useful indicator to examine the process. Thus, this method is used to analyze the hydrogen storage process. For the hydrogen compression the compression conversion coefficient amounts to $\eta_{cc} = (1207kJ/kg - 839kJ/kg)/1207kJ/kg = 247.3/1207 = 31\ \%$. When compressing air under the same conditions a lower compression conversion coefficient of $\eta_{cc}(Air) = (82kJ/kg - 61kJ/kg)/82kJ/kg = 26\ \%$ results, which emphasizes that the molar mass is not the only factor. This result is obtained because the lines of equal temperature (isotherms) in the h-s diagram of hydrogen increase with lower entropy down to temperatures of $T_{inversion(H_2)} = 193K$, while this temperature for air is around $T_{inversion(Air)} = 723K$.

For hydrogen or others fuel compression it is usual to express the compression efficiency as the ratio of technical work needed for compression to lower calorific value [Pellow et al., 2015]. For such a definition compression efficiencies of 80 % or higher can be found. However, in the context of electrical energy storage such a definition can be misleading and will not be used.

5.3 Hydrogen Combustion as Natural Gas Substitute

To adapt the Huntorf CAES machinery to hydrogen combustion the entire combustion chamber and turbine design has to be revised in order to safeguard stable and NO_x free combustion [York et al., 2015]. Such a redesign is beyond the scope of this work. However, some general considerations provide insight on the amount of hydrogen needed to energetically replace natural gas and several fuel combustion properties are discussed.

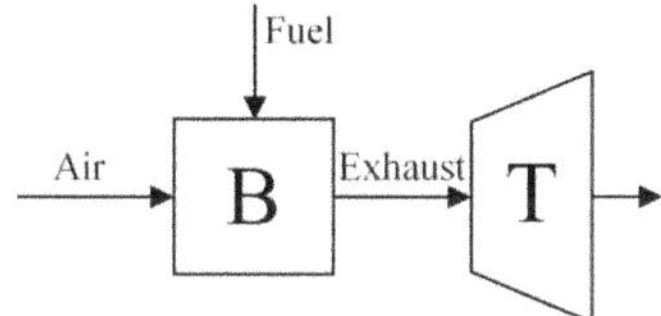

FIGURE 5.4: Process flow diagram of the combustion

Energetic equivalent amount of fuel. The required heat flow rate to heat the stored compressed air in the current Huntorf process is given by $\dot{Q} = \dot{m}_{air}\Delta h = 536.9MW$ (with $\dot{m}_{air} = 455\ kg/s$, $\Delta h = 1.18\ MJ/kg$). This heat flow rate is equal to the fuel energy $\dot{Q} = \dot{m}_{fuel} \cdot LCV$ if burner losses are negligible. For the full load steady state operation of the Huntorf plant a mass flow rate of $13.2\ kg/s$ natural gas and a volume flow rate of $16.7\ m_N^3/s$ natural gas result. The theoretically required mass flow rate of hydrogen that supplies the same amount of thermal energy is $4.5\ kg/s$ only (which corresponds in the above example to a required compressor power of $5.4\ MW$). The volume flow rate on the other hand is higher due to the low molar mass (M) of hydrogen and amounts to $50.4\ m_N^3/s$ (see Table 5.5).

Exchangeability of gaseous fuels - Wobbe index. An indicator to evaluate the exchangeability of gaseous fuels is the upper and lower Wobbe index [DVGW Deutsche Vereinigung des Gas- und Wasserfaches e. V., Mai 2008; Zachariah-Wolff, Egyedi, and Hemmes, 2007]

TABLE 5.5: Flow rates of fuel (natural gas or hydrogen) required at Huntorf process (with $\dot{m}_{air} = 455\ kg/s, \Delta h = 1.18\ MJ/kg$) and some characteristic values

	Unit	Natural gas	H_2
Relative fuel mass flow rate $\dot{m}_{fuel}/\dot{m}_{air}$	kg_{fuel}/kg_{air}	0.0291	0.0098
Mass flow rate $\dot{m}_{fuel}$	kg_{fuel}/s	13.2	4.5
Volume flow rate $\dot{V}_{fuel}$	m^3_{fuel}/s	16.7	54.6
Wobbe index (lower) W	MJ/m^3	40.9	41.0
relative density ρ_i/ρ_{air}	kg_{fuel}/kg_{air}	0.6131	0.0637

that is defined as ratio of gross calorific value (GCV) or lower calorific value (LCV) to the square root of the relative density. In this thesis the LCV is used as a reference. Thus, the lower Wobbe index W is:

$$W = \frac{LCV}{\sqrt{\rho_i/\rho_{air}}} \tag{5.1}$$

By comparing not only the heating value but also the relative density of a fuel (ρ_i/ρ_{air} with both densities at the same state conditions) similar Wobbe values indicate that the thermal load of the burner during combustion is comparable. In the example at hand (natural gas and hydrogen) it is found that the Wobbe indices are similar: $40.9 MJ/m^3$ for natural gas and $41.0 MJ/m^3$ for hydrogen, see Table 5.5 and Fig 5.5.

It is noteworthy that for a mixture of natural gas and hydrogen the Wobbe index changes over the gas composition (see Fig. 5.5). When considering the Wobbe bands (ranges of acceptable Wobbe indices) of natural gas properties according to the DVGW specifications 'Group L' [DVGW Deutsche Vereinigung des Gas- und Wasserfaches e. V., Mai 2008] that apply to the Huntorf plant (lower and upper limits of the Wobbe index 37.8 to $46.8 MJ/m^3_N$ as dashed lines in Fig. 5.5), it appears that natural gas mixtures with less than 35 % hydrogen would be allowable and that also almost pure hydrogen (more than 93.5% of H_2) also fulfills the required Wobbe index, see Fig. 5.5.

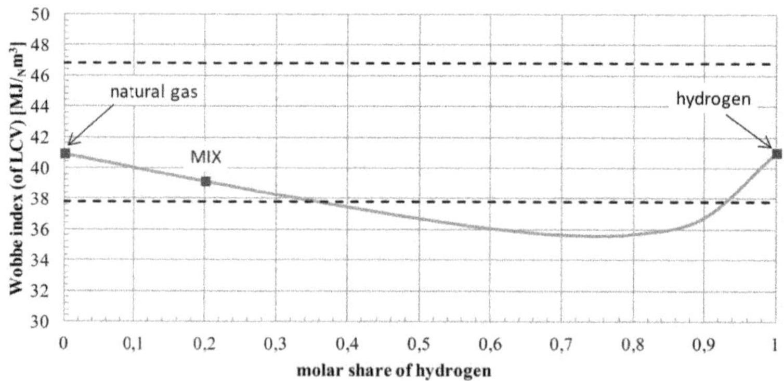

FIGURE 5.5: Wobbe index for a mixture of natural gas and hydrogen

However, this single indicator is not sufficient since other flame characteristics may determine the acceptability of the replacement gas, such as flame speed (high flame speed of hydrogen may lead to flashbacks [Zachariah-Wolff, Egyedi, and Hemmes, 2007]), flame temperature, explosion limits, completeness of combustion or composition of the exhaust gas that expands in the subsequent turbine.

Temperature of adiabatic combustion. In general the adiabatic flame temperature (T_{ad}) of hydrogen is higher than natural gas or methane adiabatic flame temperatures. However, this value also depends on the surplus of combustion air and the temperatures of air and fuel. To calculate the adiabatic flame temperature an energy balance over the combustion chamber with $\dot{Q} = 0$ for adiabatic conditions gives: $\dot{m}_{fuel} \cdot (LCV + c_p \cdot (T_{in} - T_0)) + \dot{m}_{air} c_p \cdot (T_{in} - T_0) = \dot{m}_{exh} c_p \cdot (T_{out} - T_0) + \dot{Q}$

This gives
$T_{ad} = (\dot{m}_{fuel}(LCV + c_p \cdot (T_{in} - T_0)) + \dot{m}_{air} c_p \cdot (T_{in} - T_0) + \dot{m}_{exh} c_p T_0)/(\dot{m}_{exh} c_p)$

By using the mass balance $\dot{m}_{exh} = \dot{m}_{fuel} + \dot{m}_{air}$ together with $\dot{m}_{fuel} = 1$ and $L_{min} \cdot \lambda = \dot{m}_{air}/\dot{m}_{fuel}$ it can be found that

$$T_{ad} = \frac{LCV + c_{p,fuel} \cdot (T_{in} - T_0) + \lambda \cdot L_{min} \cdot c_{p,air} \cdot (T_{in} - T_0)}{1 + \lambda \cdot L_{min}} + T_0 \qquad (5.2)$$

In the example of the Huntorf CAES plant the combustion air is 150 % higher than the stoichiometrically required combustion air flow, i.e. air ratio $\lambda = 2.5$; the temperatures of air and fuel are assumed to have a temperature of 323 K. The adiabatic temperature of different fuel compositions (mixtures with increasing hydrogen share from natural gas to pure hydrogen) is rising as depicted in Fig. 5.6.

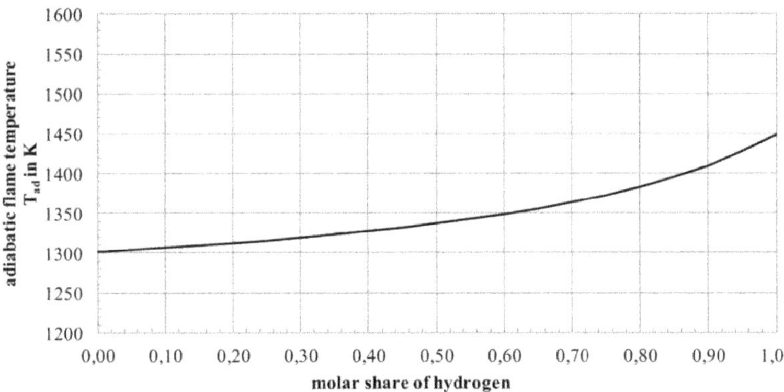

FIGURE 5.6: Adiabatic flame temperature (T_{ad}) at constant air ratio $\lambda = 2.5$ and $T_{in} = 323\ K$ and variable fuel composition

To keep the adiabatic combustion temperature (T_{ad}) constant at $1300\ K$ the air ratio (λ) has to be variable. For the example at hand an iterative solution with $\lambda(x_{H2}) = 2.5 + x_{H2}^2/2$ applies, Fig.5.7.

Exhaust gas composition and amount. The exhaust gas of the combustion units drives the turbine and is, thus, a key element determining the efficiency of the system. When changing the fuel composition, exhaust gas composition is inherently changing. Yet, for the Huntorf

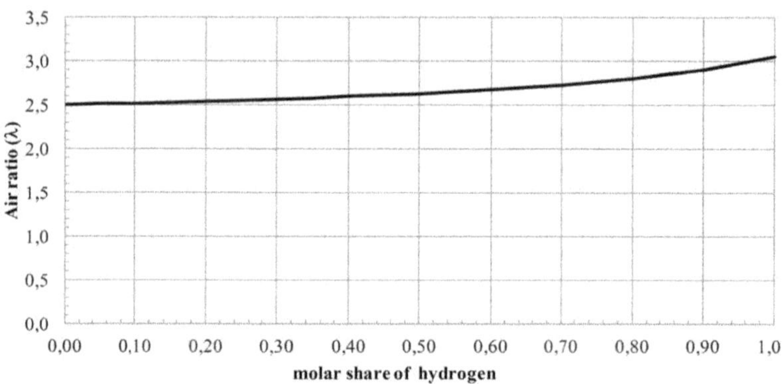

FIGURE 5.7: Variation of air ratio (λ) to keep the adiabatic combustion temperature (T_{ad}) constant at approx. $1300K$

example the effect on exhaust gas composition is relatively small since the amount of surplus combustion air is high, see Fig.5.8.

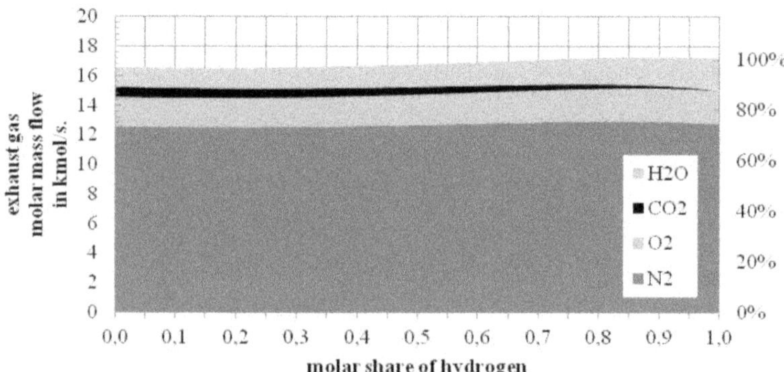

FIGURE 5.8: Molar mass flow rate of exhaust gas and its composition for rising share of hydrogen and increasing air ratio ($\lambda(T_{ad} = const.)$, see previous Figure)

Conclusion on H_2 combustion. By entirely replacing natural gas with hydrogen only a mass flow rate of one third of the fuel amount (4.4 kg/s H_2 instead of 13.2 kg/s natural gas) is needed to achieve the same heat flow rate ($\dot{Q}$ = 536.5 MW); Furthermore, a similar adiabatic flame temperature can be attained by increasing the air ratio to a value of $\lambda = 3$ for pure hydrogen (instead of 2.5 with natural gas); Wobbe indices of natural gas and hydrogen are similar and the exhaust gas mass flow rate decreases slightly. Thus, these characteristics suggest that the fuel replacement is rather unproblematic. However, the volume flow rate of the fuel triples which may affect the overall burner design considerably. Furthermore, flame velocities and flammability have to be analyzed in detail in order to safeguard a stable and NO_x free combustion as well as a low H_2 flux. In conclusion, in a first attempt to investigate

TABLE 5.6: Natural gas or hydrogen flow rates, air ratio (λ) and exhaust gas flow rates for a fixed air mass flow rate, heat flow rate ($\dot{Q}$) and adiabatic combustion temperature (T_{ad})

Air	$\dot{m}$	kg/s	455	
	$\dot{n}$	$kmol/s$	15.777	
	$\dot{V}_N$	m_N^3/s	353.6	
Heat flow rate	$\dot{Q}$	MW	536.5	
Adiabatic combustion temperature	T_{ad}	K	1281	1270
Fuel			NG	H_2
	$\dot{m}$	kg/s	13.2	4.44
	$\dot{n}$	$kmol/s$	0.748	2.22
	$\dot{V}_N$	m_N^3/s	16.8	49.7
	λ	-	2.5	3.0
Exhaust gas	$\dot{m}$	kg/s	468.2	459.4
	$\dot{n}$	$kmol/s$	16.528	16.882
	$\dot{V}_N$	m_N^3/s	370.4	378.4

a hydrogen driven Huntorf process one can assume that with an adapted burner design such a process is feasible and that a fuel replacement with the above stated process characteristics is possible.

5.4 Discussion

In the previous sections, several factors were presented that have to be considered when substituting natural gas by hydrogen. While the actual storage underground is already proven in operation, the high effort to compress hydrogen has to be considered: Due to the low molecular weight of hydrogen the compression of 1 kg hydrogen is more than 14-times more costly (in terms of specific technical work) then a similar compression of 1 kg air. On the other hand higher compression conversion coefficients are applicable due to the low inversion temperature of hydrogen. The use as fuel in combination with gas turbine technology still requires research and development. The estimate in the previous section present the basic idea of such a concept. However, such general considerations are not sufficient to confirm the feasibility. More detailed investigation including experimental validation is required before an implementation of such options becomes feasible.

Chapter 6

Economics of CAES

Profitable operation is a key feature of any new technology. To investigate the potential profits of CAES, one has to consider the energy market conditions together with the characteristics of the CAES system. Several such studies have previously been performed for CAES concepts in the German energy market conditions by *ex-post* analysis of different historical market price data sets: [Wolf, 2011] (for the year 2007), [Madlener and Latz, 2013] (2007), [Kaldemeyer, Boysen, and Tuschy, 2016] (2012-2014) and [Wirth, 2015-08-03] (2014-2015).

Madlener and Latz [2013], Kaldemeyer, Boysen, and Tuschy [2016], and Wirth [2015-08-03] compare adiabatic and fuel-driven CAES concepts and find that fuel-driven CAES plants achieve the highest profits in the investigated German market frame set. However, they [Madlener and Latz, 2013; Kaldemeyer, Boysen, and Tuschy, 2016] argue in favor of adiabatic CAES to avoid the carbon dioxide (CO_2) emission of CAES plants with carbon-based fuels, since such fuel-driven CAES plants are not in accordance with the fundamental idea of the transition to renewable (greenhouse gas neutral) energies [Madlener and Latz, 2013; Kaldemeyer, Boysen, and Tuschy, 2016]. In the future, a changed regulatory frame-work of the energy markets might penalize CAES plants with carbon-based fuels, for instance, in the form of higher prices for CO_2-certificates.

The CAES power design in such studies is a determining factor of the profitability. Madlener and Latz [2013] used a compressor to turbine power ratio of 1:2 based on assumptions presented in [Gatzen, 2008]. Kaldemeyer, Boysen, and Tuschy [2016] chose a 1:1 ratio for their economic studies. On the other hand, the extensive optimization model used by Wolf [2011] found that a compressor to turbine power ratio of 7:4 is economically ideal when acting on spot and tertiary reserve markets (*ex-post* optimization for the year 2007 [Wolf, 2011]. This indicates that low-price periods from the market are rather short. Thus, compression power has to be high in order to optimally utilize these short low-price periods for loading the storage via air compression. Periods with higher prices, on the other hand, have a longer duration so that turbine power can be comparatively low.

Energy markets. The German energy market can be subdivided into two: energy trade and grid services. Both types are composed of a multitude of different trading options and products. These are, respectively:

- trade at stock exchange (e.g. day ahead or intra-day) or 'over the counter' and
- grid services such as primary and secondary control, tertiary reserve, reactive power compensation, black start ability or grid reserve.

Currently, the most relevant energy markets for CAES products are the spot and tertiary reserve markets as well as other electrical grid ancillary services (reactive power compensation, black start ability or grid reserve) [Wirth, 2015-08-03]. A combination of both, the tertiary reserve (also called minute reserve) and the spot market, is found to be most profitable in *ex-post* analysis based on historical data [Wolf, 2011; Madlener and Latz, 2013; Kaldemeyer, Boysen, and Tuschy, 2016] with the tertiary reserve market having the largest share of the profits [Wolf, 2011; Madlener and Latz, 2013; Kaldemeyer, Boysen, and Tuschy, 2016; Wirth, 2015-08-03].

Fig.6.1 shows the electric energy load of Germany in a particular week in December 2016 [*Energy Charts: Preise*]. The contribution of renewable energy sources (solar and wind) as well as conventional power plants and the electric energy import balance is depicted. Electric energy stock exchange prices for every 15 minutes are displayed as overlaying graphs. Some negative energy prices can be observed, which often occurred on weekends.

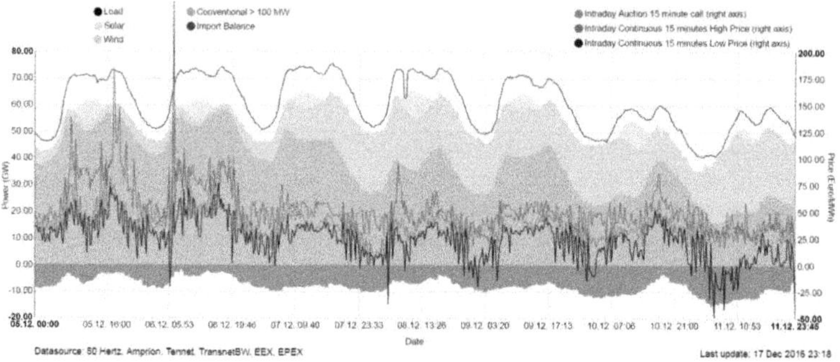

FIGURE 6.1: Electricity production and spot prices in week 49 of 2016 [*Energy Charts: Preise*]

Profit of CAES systems. The profit of a CAES system is determined by the revenues reduced by the fixed and the variable costs. In all the previous studies [Wolf, 2011; Madlener and Latz, 2013; Kaldemeyer, Boysen, and Tuschy, 2016; Wirth, 2015-08-03], the largest revenues are seen to be obtained from the tertiary reserve market, followed by spot market trading. However, Kaldemeyer, Boysen, and Tuschy [2016] demonstrate that the absolute values of revenues vary considerably in different reference years (2012-2014 [Kaldemeyer, Boysen, and Tuschy, 2016]). Wirth [2015-08-03] presents the revenues from other electric grid ancillary services of the Huntorf plant that can be considered as a "fixed revenue" due to long-term agreements and little to no impact on operational schedules. Variable costs mainly consist of energy costs (for electric energy to run the compressors, fuel cost and CO_2 certificates) as well as the maintenance cost. The fixed cost is mainly the capital cost.

Capital cost. The capital cost of a CAES plant is composed of the cost for the machinery (generator, turbine, compressor, throttle, piping and cooling apparatus), the air storage system, and the balance of the plant (BOP) (buildings, roads, electrical systems, pollution control and others). While BOP applies equally to any other power plant, the cost for air storage is an additional cost only applicable to CAES plants. However, this extra cost is compensated by the fact that for the same machinery size, a higher output power can be obtained from a CAES plant (with a factor of 2.5 compared to a conventional gas turbine) which directly decreases specific cost.

The capital cost figures from the existing CAES plants Huntorf and McIntosh are around 600 to 750$€_{2017}/kW_{el}$ (of turbine power) [EPRI and the U.S. Department of Energy, 2003].[1]

The specific cost per storage volume of a solution-mined salt cavern is site dependent and approximately 39 to 57 $€_{2017}/m^3$, i.e. for a $500,000\ m^3$ salt cavern, 19.5 to 28.5 Million € [Fichtner GmbH, 2014], (p.118, Fig. 4-31). For a salt cavern of the $500,000\ m^3$ size or larger, almost no cost digression applies, so, a storage volume of $1\ Mm^3$ split in two caverns has approximately the same specific cost (making a total cost of around 40 to 56 Million $€_{2017}$).

The range of different cost estimates from different sources is summarized in Table 6.1. Figures from Bailie [2017-07-12] are rather high and will be used as a conservative estimate. However, in favorable conditions, lower costs are applicable. As a calculation example, a CAES plant with a capital cost of 320 MW with an air storage volume of $300,000\ m^3$ (which corresponds to the Huntorf CAES dimensions) is estimated to cost: $1050\ €/kW^{[2]} \cdot 320\ MW + 52\ €/m^3 \cdot 300,000\ m^3 = 352\ M€_{2017}$ or a specific cost of 1100 $€/kW$. However, the actual cost of the Huntorf CAES plant were only half of this amount (a total cost of 171 M $€_{2017}$ and specific cost of 590 $€/kW$ [EPRI and the U.S. Department of Energy, 2003]).

TABLE 6.1: Cost of CAES

	size	total cost M $€_{2017}$	specific cost $€_{2017}/kW$	reference
Huntorf 1978	290 MW_{el}	171	590	[EPRI, 2003]
McIntosh 1991	110 MW_{el}	[a] 84	766	[EPRI, 2003]
Apex project 2013	317 MW_{el}	[b]193	609	[Dresser-Rand, 2013]
Literature values				
			560-870	[BMWi, 2014]
			420-840	[Chen et al., 2009]
			[c] 420	[Greenblatt et al., 2007]
			[d] 500-600	[Bailie, 2017-07-12]
			[e] 890-940	[Biasi, 2009]
Example				
- machine	320 MW_{el}	176	550	[Bailie, 2017-07-12]
- bop	320 MW_{el}	160	500	[Bailie, 2017-07-12]
- cavern	300 000 m^3	16	52 $€_{2017}/m^3$	[Fichtner GmbH, 2014]
- total		352	1100	

[a] In 1991 the total cost of the McIntosh CAES plant was 51 Mio. $\$_{1991}$ [EPRI and the U.S. Department of Energy, 2003]. The U.S. dollar experienced an average inflation rate of 2.28 % per year between 1991 and 2017 [*Inflation calculator*]. $51 in the year 1991 is worth $91.56 in 2017 [*Inflation calculator*]. The exchange rate of 0.9205 from $ to € (March 2017) is used.

[b] Press release [Dresser-Rand Group Inc., 2013]

[c] only CAES expander and compressor

[d] machinery without bop or storage

[e] including storage

[1] Using inflation calculation to transform the 2002 figures to 2017, where the U.S. dollar experienced an average inflation rate of 2.04 % per year between 2002 and 2017 [*Inflation calculator*]; and the currency exchange rate of $1 = €0,9205 in March 2017.

[2] Based on cost estimates for machinery (550 €) and bop (500 €) according to [Bailie, 2017-07-12] which is in good agreement with estimates from Kaldemeyer, Boysen, and Tuschy [2016] of approx. 500 $€_{2017}$ for the machinery plus an additional 50 % of the total cost for bop.

Chapter 7

Case Study of 100 % Renewable Lower Saxony with CAES

The above developed calculation methods (Chapters 3, 4 and 5) are used to estimate the process characteristics of the next generation CAES design. This system is used in the case study of a 100 % renewable energy system in the Lower Saxony region of Germany.

7.1 100 % Renewable Energy Lower Saxony

Faulstich et al. [2016b] developed a back casting scenario for a future 100 % renewable energy-powered Lower Saxony region. The assumptions are based on available surface areas and resources (e.g. how many roofs are available for PV solar power and how many woods can contribute to biomass production, respectively) as well as estimates of future energy loads in Lower Saxony (mainly based on population and economy growth). This region represents approximately one-tenth of Germany's surface area as well as one-tenth of its population. The electric energy, heating, industry, and mobility sectors are considered. Hence, this energy system model is highly interconnected as illustrated by Fig.7.1. It has been discussed with stake holders from politics, society, and industry in order to create a realistic scenario [Faulstich et al., 2016b]. However, several future developments, such as economic and population growth, cannot be foreseen. Thus, a large uncertainty remains.

Using assumptions developed in [Faulstich et al., 2016b], Brendel and Niepelt [2016] estimate future renewable energy production and future loads on an hourly basis [Faulstich et al., 2016a]. Wind and solar power (as the most important renewable energy sources in Lower Saxony) are estimated based on historical weather data through five reference years, 2011 to 2015. Assumptions for installed power capacities are calculated based on spacial potential and stakeholder constraints. For Lower Saxony, 53 MW wind power and 89 MW PV are the values estimated for a 100 % renewable energy-powered region in the future. Certain renewable energy sources, such as geothermal energy, are negligible and will not be further discussed. Electric energy load is estimated through assumptions of demographic developments and advances in energy efficiencies, and it fluctuates around a mean value of 11.4 GW (in a 8.1 to 14.3 MW range). A random sample week is displayed in Fig.7.2 to illustrate these figures. In contrast to the current electric grid situation (Fig.6.1), no import or export of electricity is allowed in order to prove the feasibility of a self-sustaining energy system based on renewables alone, without conventional power plants in this specific region. The installed capacities of wind and PV power plants exceed the maximum load by far, see Fig.7.2 and Table 7.1. However, at night, energy conversion from renewables is often quite

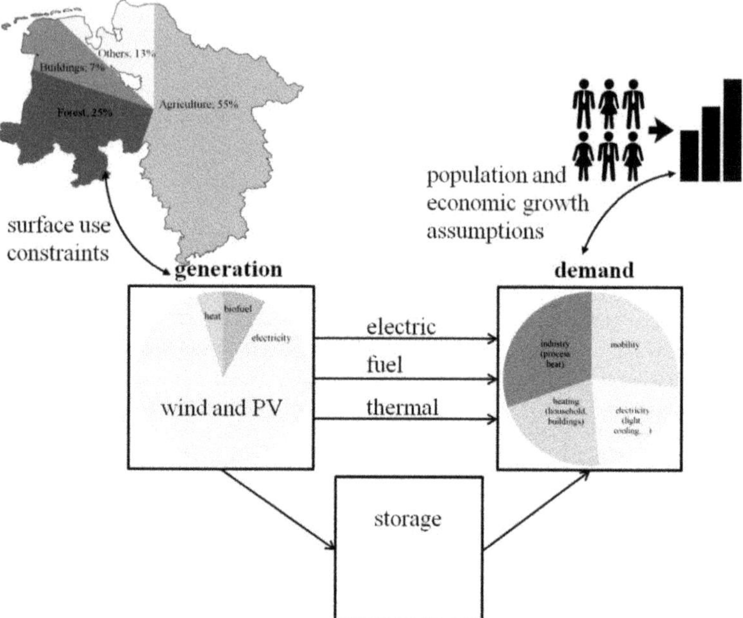

FIGURE 7.1: Lower Saxony 100 % renewable energy system

low (e.g. in Fig.7.2, on March 19th to 22nd, a deficit of around $70\ GWh$ occurs during the night). Thus, energy storage is required to balance electric energy generation and demand. A detailed description of the assumptions for the scenario is provided and discussed in [Faulstich et al., 2016b; Faulstich et al., 2016a; Brendel and Niepelt, 2016].

The aim of this Chapter 7 is not re-calculating the overall system model, but a discussion of the potential role of CAES, which was not considered in the original scenarios, in such a 100 % renewable energy system.

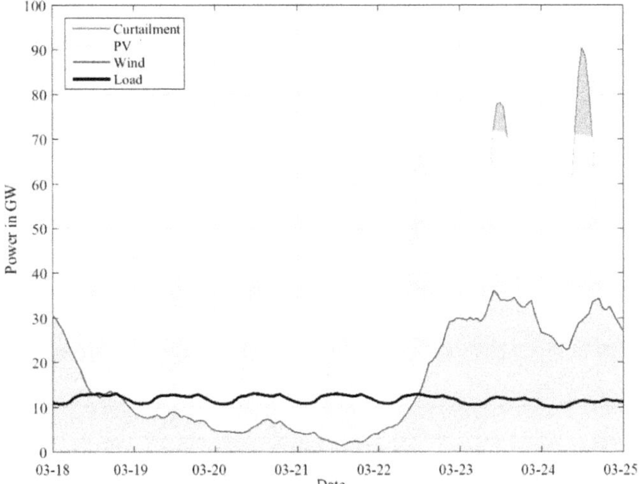

FIGURE 7.2: Electric energy profile in a 100 % renewable Lower Saxony, data adapted from [Brendel and Niepelt, 2016]

Operation schedule of energy storage. Energy storage systems are required to balance fluctuating renewable energy generation and loads (Fig.7.2). Faulstich et al. [2016a] investigate three types of storage: PHES, Li-ion batteries and chemical energy storage via hydrogen (water electrolysis, hydrogen storage and subsequent combined cycle hydrogen turbines). The following storage characteristics are considered by Brendel and Niepelt [2016]:

- Li-ion: $\eta = 84.9\ \%$; power range -27 MW to +23 MW
- PHES: $\eta = 80.1\ \%$; power range -188 MW to +188 MW
- H_2-path: $\eta = 37.8\ \%$: P_{el}(Electrolysis)= -47.3 GW; P_{el}(H_2-Turbine) = +12.5 GW

The yearly electric energy consumption amounts to $99.96\ TWh_{el}/a$ based on a five year average. The average wind and PV electric energy generation amounts to $204.1\ TWh_{el}/a$, i.e. more than the double load, see Table 7.1. However, only $83.6\ TWh/a$ of this renewable energy is produced at 'the right time'. Excess renewable energy is stored (mainly via electrolysis), transferred to the heating sector (via heat pumps) or wasted via curtailment (Fig.6.1 on March 23rd and 24th). The deficits in the electric grid are balanced with hydrogen turbines at a

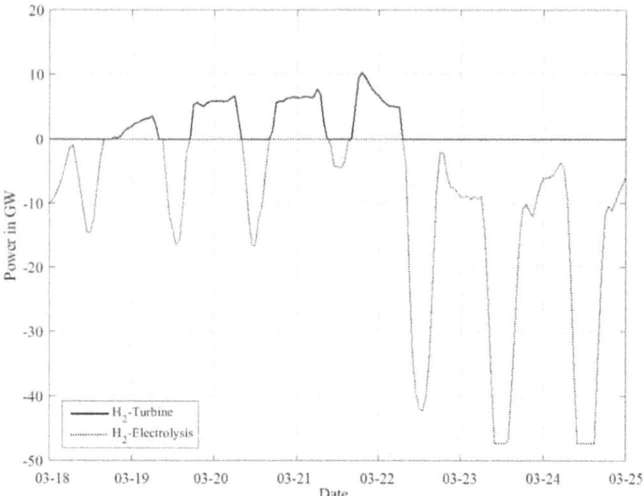

FIGURE 7.3: Operation times of electric energy storage in a 100 % renewable Lower Saxony, data adapted from [Brendel and Niepelt, 2016]

TABLE 7.1: Average electric energy generation in the Lower Saxony case study (without CAES)

	Wind	PV	H_2 turbine	H_2 electrolysis	load
Power in GW	53.0	89.0	12.5	-47.3	-11.4 (av.)
Energy in TWh_{el}/a	95.1	109.0	16.4	-78.6	-99.96
Hydrogen TWh_{H2}/a			-28.1	69.1	

maximum power of 12.5 GW, which is sufficient capacity to satisfy the average electrical load. The storage capacity amounts to 18.8 TWh of H_2 storage capacity.

Due to economic reasons, Faulstich et al. [2016a] used a negligible share of pumped hydro and battery energy storage units. Thus, the residual load was almost entirely served by the hydrogen storage path. The surplus of renewable energy production is transformed via water electrolysis into hydrogen (H_2). When a shortfall of electric energy occurs, the stored hydrogen is turned back into electricity via turbines. The operation times depicted in Fig.7.3 correspond to the residual load shown in Fig.7.2. This storage process comes along with its corresponding energy losses. It is important to note that parts of the produced H_2 are used in industrial processes or in the mobility sector. Thus, hydrogen electrolysis not only serves electric energy storage but also interlinks electricity to the industry and mobility sectors. Furthermore, heat from the hydrogen turbines is used in the heating sector.

The dynamic operation schedule of the next generation H_2 CAES in a future energy system is based on the operational characteristics of water electrolysis and hydrogen turbines found by Brendel and Niepelt [2016]. This entails a changed efficiency regime for the overall energy system as well as a different proportion of energy technology capacities.

7.2 Next Generation CAES Concept for Renewable Energies

The characteristics of a carbon free hydrogen driven CAES concept, based on realistic assumptions found in the previous chapters, are analyzed in a 100 % renewable environment model. All properties are estimated based on those of the existing Huntorf and McIntosh plant. Efficiency values are chosen in a conservative 1 % range better than those of the existing plants with respect to advances in technology. The McIntosh design with exhaust gas enthalpy recuperation is used to reduce exhaust gas enthalpy losses and, thus, achieve higher efficiencies. A four-stage compressor train and a two-stage expansion is chosen. Salt caverns for air and hydrogen storage are assumed. The system properties are summarized in Table 7.2.

TABLE 7.2: Assumptions for the case study

Property	Value	Unit
Fuel	H_2	
Power $P(Turbine)$	520	MW
$P(Compressor)$	385	MW
Air storage cavern size	10^6	m^3
Turbine start up duration	11	min
Compressor start up	4.5	min
$p_{max}(Cavern)$	76	bar
$p_{max}(Turbine)$	42	bar
$T_{max}(Turbine)$	1250	K
$\eta_{mech,C}$	0.94	
$\eta_{mech,T}$	0.98	
$\eta_{s,C}$ per compression stage	0.91, 0.85, 0.82, 0.82	
$\eta_{s,T}$	0.90	

7.2.1 Steady State Properties

Based on the assumptions delineated in Table 7.2, the steady state properties of the next generation CAES can be estimated. Fig.7.4 shows the h-s diagram at a maximum cavern pressure of 76 bar. From these calculations, the technical work of compression can be determined; $w_{t,c} = 520\ kJ/kg$ (at 76 bar) or $w_{t,c} = 440\ kJ/kg$ (at 42 bar). The technical work of the turbine is kept constant at a value of $w_{t,t} = 805\ kJ/kg$ due to throttling to a turbine inlet pressure of 42 bar. The mass flow rates in compression and turbine mode are $\dot{m}_c = 700\ kg/s$ and $\dot{m}_t = 650\ kg/s$. Based on these steady state calculations the energy efficiency values are calculated, see Table 7.3.

η_{caes} %	η_{th} %	η_{rt1} %	η_{rt2} %	η_{rt3} %	η_{rt4} %	hr_1 $\frac{kWh_{fuel}}{kWh_{electric}}$	hr_2
55	30	155	90	82	4.8	3.3	1.2

TABLE 7.3: Comparison of the different energy efficiencies and heat rates of the next generation CAES

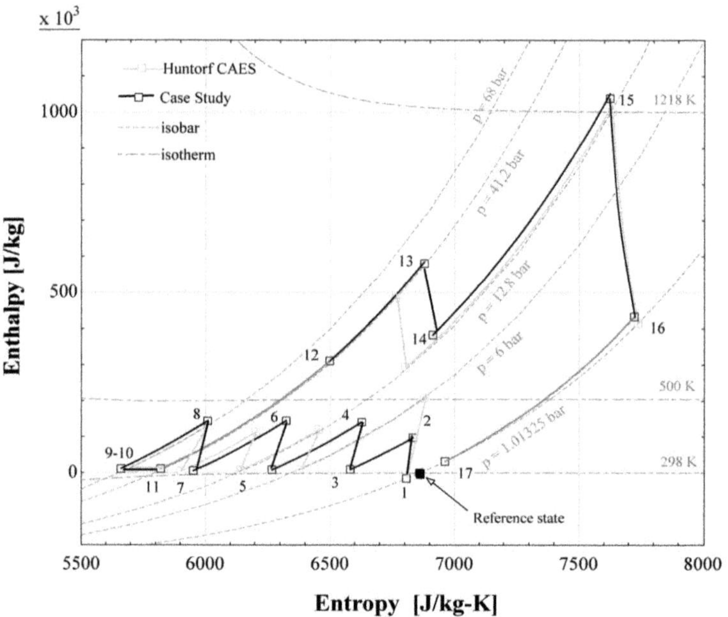

FIGURE 7.4: h-s diagram of the case study next generation CAES process (black) in comparison with the current Huntorf CAES process (in grey)

7.2.2 Dynamic Operation

The dynamic operation characteristics of CAES as examined in Chapter 4 are now applied to the case study. These are as follows:

- Technical work for compression rises with rising cavern pressure;
- Start-up times of turbine and compression are 11 and 4.5 minutes, respectively; other switch times are listed in Tab.4.4;
- Full load turbine operation is possible down to a certain cavern pressure; below that value, the turbine starts to operate in part load; and
- Heat rate (hr_2) and air rate (ar) vary with turbine power.

The combination of the above design and the operation schedule in the sample week of the 100 % renewable energy-powered Lower Saxony case study results in the operation pattern depicted in Fig.s 7.5 and 7.6. It can be observed that the CAES system can replace a considerable share of the formerly used H_2-turbine. However, its contribution is limited by the storage capacity. The charging level corresponds to the cavern pressure established in Fig.7.6. Its maximum of $76\ bar$ is reached on Day 1 of the sample week so that the solar peak production cannot be stored via air compression. On Day 4 the turbine starts to operate in a part load because cavern pressure falls below full-load minimum pressure of $42\ bar$. On Day 5, the part-load operation continues until the operational minimum of $20\ bar$ cavern pressure is reached. The remaining residual loads have to be served by the H_2-turbine.

In the sample week the cavern air temperature varies over time in a $290\ K$ to $350\ K$ range, Fig.7.6.

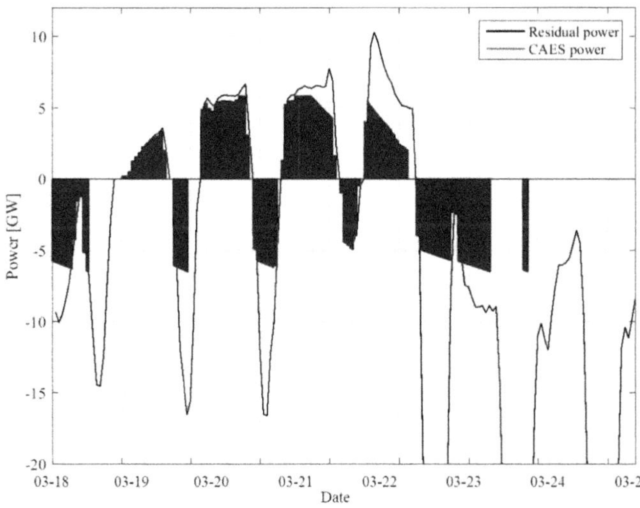

FIGURE 7.5: CAES Power in the sample week

TABLE 7.4: Average electric energy generation in the Lower Saxony case study with CAES

	H_2 turb.	H_2 electrolysis	CAES turb.	CAES comp.
Power in GW_{el}	6.25	-42.3	6.25	-5
Energy in TWh_{el}/a	8.2	-72.9	8.2	-6.1
Hydrogen in TWh_{H2}/a	-14.0	64.2	-9.1	

In the adapted 100 % renewable energy Lower Saxony scenario, the capacities of the hydrogen path are partly replaced with CAES capacities, see Table 7.4. The better turbine conversion efficiency η_{tc} of CAES entails that a theoretical maximum of 4.9 TWh less hydrogen has to be produced and stored.Hence the corresponding losses are avoided. However, due to the limited storage capacity of CAES, the actual amount of avoided hydrogen production is lower. Furthermore, the CAES compressor power replaces some of the required electrolysis power but helps to raise full-load working hours of the electrolysis.

7.3 Results

In the dynamic plant operation (which comprises rising or falling cavern pressures according to the charging level of the cavern, part load operation as well as start-up procedures), the actual efficiency values are considerably lower than those found using steady-state assumptions. For the sample week depicted in Fig. 7.5, or more precisely for the overall discharging and charging period marked in Fig. 7.6 the CAES efficiency η_{caes} falls to 47.5 % compared to the 55 % found using steady-state assumptions. This value is calculated using the actual electrical energy provided by the system $W_{el,G} = 197.7\ GWh$ and the actual energy consumption of $W_{el,M} = -173.3\ GWh$. Further, the actual heat rate (hr) as a function of turbine power has been calculated with a nameplate value of 1.2. Hence, the actual hr value is higher if a part load operation is required. The overall fuel required amounts to $243.12GWh$ in the storage cycle of the sample week. Thus, an actual thermal efficiency of $\eta_{th} = 10.1$ % is obtained in contrast to the steady state value of 30 %. The ratio of electric energy production to consumption amounts to $\eta_{rt1} = 1.14$, which is also considerably lower than the steady state value of 1.55.

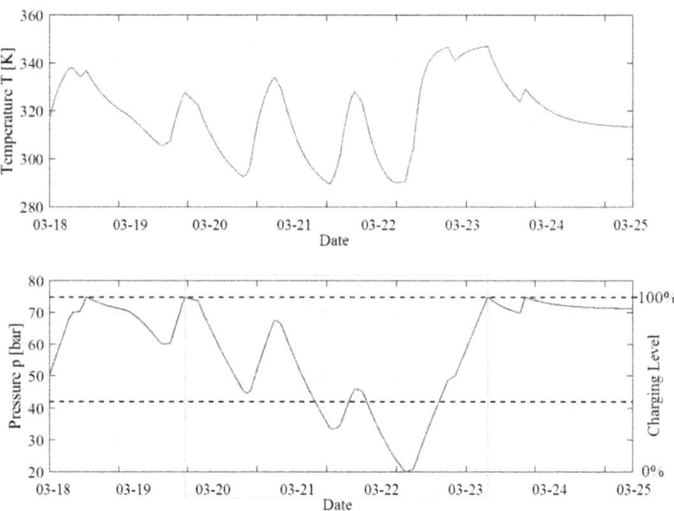

FIGURE 7.6: Temperature and pressure of the air storage place for the sample week

Chapter 8

Conclusions

In this doctoral thesis, a detailed steady state and time-dependent thermodynamic analysis of existing and forthcoming CAES processes is presented. The calculation methods used to determine the properties of the process are validated with a unique set of measured operational data of the Huntorf CAES plant. Several aspects of a hydrogen-driven CAES plant are discussed. Insights to economics of CAES are provided. These aspects are then used to present a case study of a hydrogen-driven next generation CAES system for a 100 % renewable energy system in Lower Saxony in the future. Corresponding operation patterns and characteristic values of the process are presented.

Steady state and time-dependent thermodynamics of CAES. Comparing the results of the process characteristics in steady state conditions and the time dependent operations, it is found that the resulting values differ significantly. While the effect of isobaric air storage cavern, as an inherently time-dependent process unit, is rather low with less than 1 percent point in terms of CAES efficiency at the Huntorf example (see Chapter 3), the effect of the part load operation is considerably higher. In the sample week (Chapter 6), the CAES efficiency of the next generation case study was 7.5 percent points lower than the steady state estimate. This corresponds to a relative overestimation of 14 % of the process efficiency η_{caes}. However, this example represents the worst-case scenario since only one charging cycle, including the entire part load operation range, is considered. This calculation emphasizes the importance of not only using appropriate assumptions for thermodynamic calculations (such as irreversible thermodynamics and properties of air as real gas, as discussed in Chapter 3) but also of considering the actual operational patterns (start-ups, part load, and run-down procedures).

Electric energy storage efficiency of CAES. Because of the electrical *and* fuel energy input, the definition of electrical energy storage efficiency for the CAES process is unambiguous (except for fuel-free adiabatic CAES). In Chapter 2, different approaches are discussed in detail. Hence, one can state that CAES electrical energy storage efficiency is in a 3-160 % range, depending on the definition used ('efficiencies' larger than 1 are obtained when calculating the ratio of the input and output electrical energy η_{rt1} without considering the energy contribution of the fuel, see Eq.2.11). Hence, any efficiency value has to be supplemented by the appropriate calculation method. Comparisons between strictly electric energy storage systems (such as pumped hydro and batteries) and CAES cannot be based on a single efficiency value, but requires more comprehensive approaches.

The round-trip efficiency η_{rt4} is ideal in tracing the conversion of electrical energy input through the system. It can be split into a compression and a turbine conversion efficiency: $\eta_{rt4} = \eta_{cc} \cdot \eta_{tc}$, where the compressor conversion efficiency is defined as $\eta_{cc} = \frac{E_{air}}{W_{el,M}}$. In the

Huntorf and McIntosh examples, it is around $\eta_{cc} = 5$ %, i.e. 95 % of the electric energy input is wasted as heat. The turbine conversion coefficient is $\eta_{tc} = \frac{W_{el,G}}{E_{air}+Q_{fuel}}$. It amounts to 57 % for Huntorf or 80 % for McIntosh CAES plants.

It then appears that the full potential of CAES is not only electric energy storage (or more precisely providing positive and negative power reserve) but also sectoral interlinking: charging consists of air compression and *power-to-heat* technology; discharging consists of the turbine operation with highly efficient (i.e. better than Carnot) combustion via compressed air usage of a renewable fuel such as green[1] hydrogen (*chemicals-to-power*).

Exergetic analysis of CAES. The benefits of the exergetic system analysis include a changed perspective on the energy content of stored compressed air at ambient air temperatures. While this energy content in terms of enthalpy is close to zero (thereby implying low efficiencies of the compression conversion process, i.e. the charging process), the *exergy* content of compressed air is considerably higher. However, for any comparison in an electric energy context, this has no further implications since electric energy is strictly exergy. Additionally, in the context of a heat engine, exergetic characteristic values have limited applicability due to the necessity of LCV-based heat energy content estimates. Thus, it appears that many exergetic analyses contributed to the highly confusing set of literature values on CAES efficiency.

Economics. In a conventional gas turbine, approximately two-thirds of the turbine power output is used to drive the compressor. The CAES turbine process runs without simultaneous compression; therefore, the power output triples at a similar turbine dimension. Hence, the power specific cost of CAES turbines must be lower. However, there are no off-the-shelf CAES power plant solutions yet, and the additional cost for air storage has to be considered, which makes CAES cost site-specific. Depending on the cost of electricity, CAES can provide much better return of investment than conventional gas power stations. Further, in comparison with pumped hydro energy storage, it appears that CAES is generally assumed to be more cost efficient.

CAES - Electric Energy Storage? Eventually, one has to conclude that the value of CAES is largely underestimated if it is solely considered as an electric energy storage system. This is a result of its dual character as a mechanical and a thermal energy storage system as well as its dual character as storage and generation units. Fuel-driven CAES offers the unique feature of linking all three energy storage sectors: electric, thermal, and chemical. Its full potential can only be estimated in the energy system analysis with an overall energy system (heat, electric, chemical) and the appropriate process characteristics.

[1] i.e. hydrogen from electrolysis driven with power from renewable energy sources

Appendix A

Measured Data of the Huntorf CAES Plant

[Krüger, 27.07.2015]

A.1 Charging

During charging the air storage cavern air mass flow rate, compressor power, temperatures (T_1 to T_9) and pressures (p_1, p_2, p_4, p_6, p_8 to p_{10}) have been measured. The time intervals of these breakpoints are not specified. Ambient temperature T_1 is around $10^oC(+/-0,5K)$ and the air mass flow is almost constant at 108 kg/s (+0,7/-0,3 kg/s). During the measurement, the cavern pressure p_{10} rises from 53 to 59 bar.

TABLE A.1: Charging 25.10.1980

Breakpoint		1	2	3	4
Air Mass Flow Rate in [kg/s]		108.5	108.5	108.7	107.7
Compressor Power in [MW]		66.8	67.6	67.6	68.5
Temperature in [deg C]	T_1	9.5	10.0	10.3	10.5
	T_2	235.1	237.8	238.5	241.1
	T_3	31.3	32.1	31.9	32.9
	T_4	148.0	149.8	149.8	152.8
	T_5	34.5	35.5	35.2	36.6
	T_6	126.4	128.2	128.6	131.3
	T_7	34.5	35.5	35.4	36.5
	T_8	144.0	147.1	148.0	151.7
	T_9	49.3	50.6	50.3	52.0
Pressure in [bar]	p_1	1.0	1.0	1.0	1.0
	p_2	6.1	6.1	6.2	6.3
	p_3	5.9	6.0	6.0	6.1
	p_4	13.9	14.2	14.3	14.7
	p_6	27.9	28.8	29.3	30.7
	p_8	57.1	59.9	62.3	66.7
	p_9	55.8	59.1	60.9	65.3
	p_{10}	52.9	55.3	56.8	58.7

A.2 Discharging

For discharging measured values of air mass flow rate, turbine power, temperatures (T_{11} to T_{16}) and pressures (p_{10}, p_{11}, p_{13}, p_{14} and p_{16}) over a time span of 8 hours are given, as well as three sets of data at different rated powers.

Mass Flow Rate and Power. At the beginning of the measurement over time (breakpoint 1 at 8h30), the mass flow rate is zero and is then increased to a maximum value of 408 kg/s (breakpoint 2 at 8h53). It follows a gap of measured data until 10h28 with 396 kg/s, hence, when assuming linear interpolation mass flow remains almost constant or slightly decreases. The mass flow rate is then cut to 218 kg/s (from 10h37 until 10h57) and afterwards increases to its former level of 397 kg/s (11h10). This value remains approximately constant until 12:00 when the mass flow starts to decrease constantly until breakpoint 21 at 16h49 to a value of 237 kg/s. Thus, the measured values are not always taken in regular time intervals but some larger and some shorter time spans occur. In the time span from 12:00 to 16:30 a linear interpolation of the value seems appropriate, but it is to question if this is applicable to the larger gap between 9:00 and 10:30. Fig. A.1 shows the air mass flow rate and the turbine power. Fig. A.2 shows the linear relation of both.

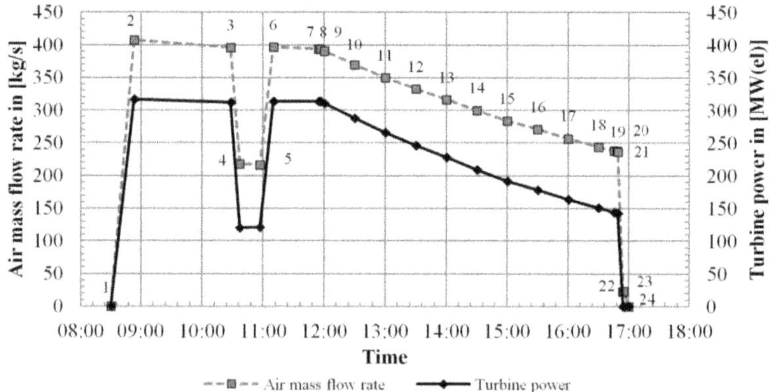

FIGURE A.1: Discharging of the air cavern – air mass flow and turbine power output over time

Temperatures and Pressures. During discharging the cavern temperatures and pressures over time have been measured. All values are presented in Fig. A.3. At the beginning and the end of the measurement, when the air mass flow rate is zero or close to zero (breakpoints 1, 23, 24), temperature T_{12} falls rapidly. Thus, heat transfer of the metering point to the surroundings is very likely.

The difference of pressure p_{10} and p_{11} corresponds to the dynamic pressure loss of the filter and is proportional to the air mass flow rate, as illustrated by Fig. A.4. In Fig. A.4 some breakpoints at the beginning (1 to 5) and the end of the measurement (21 to 24) are not depicted since they do not reflect this linear relation.

Temperatures T_{11} and T_{12} correspond to the temperature before and after throttling. Kaiser, Weber, and Krüger [2018] assume that the change of temperature during throttling

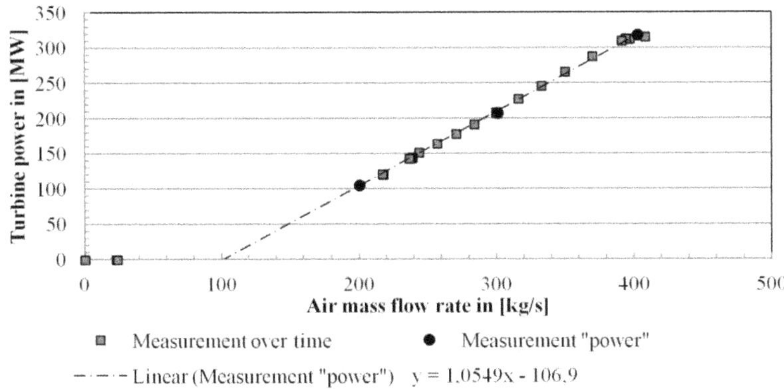

FIGURE A.2: Turbine power over air mass flow rate during discharging the compressed air storage cavern

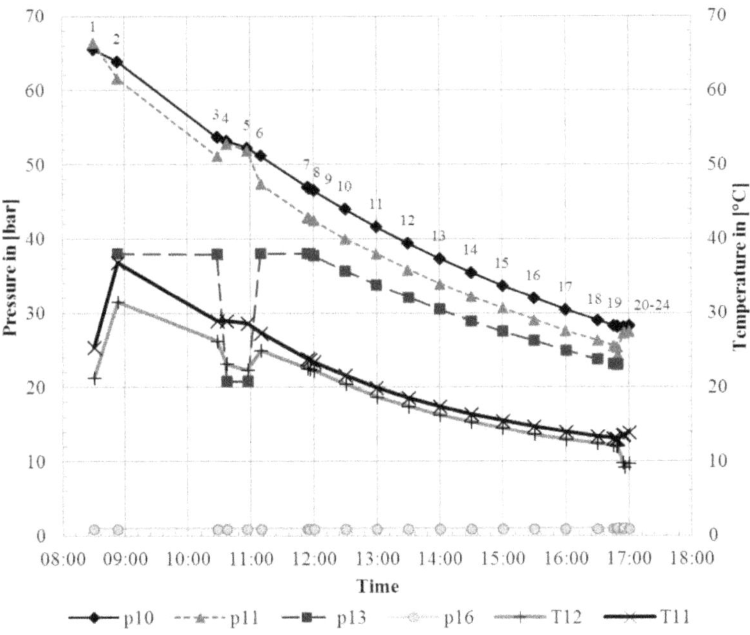

FIGURE A.3: Temperatures and pressures over time during discharging of the air cavern

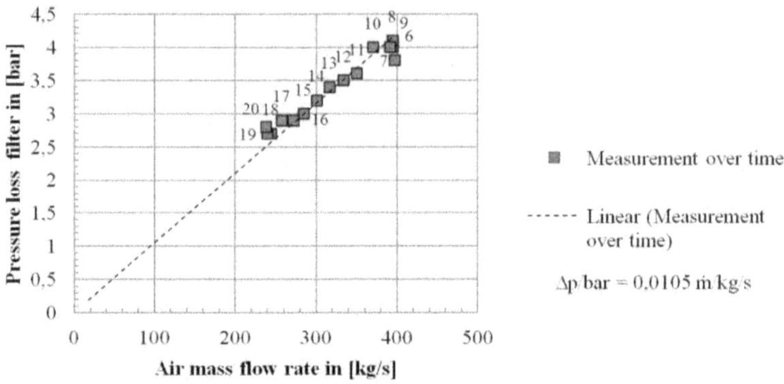

FIGURE A.4: Pressure loss of the filter ($\Delta p = p_{10} - p_{11}$) over the air mass flow rate

can be determined by the Joule-Thomson effect. Fig. A.5 confirms the linear relation of temperature difference to pressure drop.

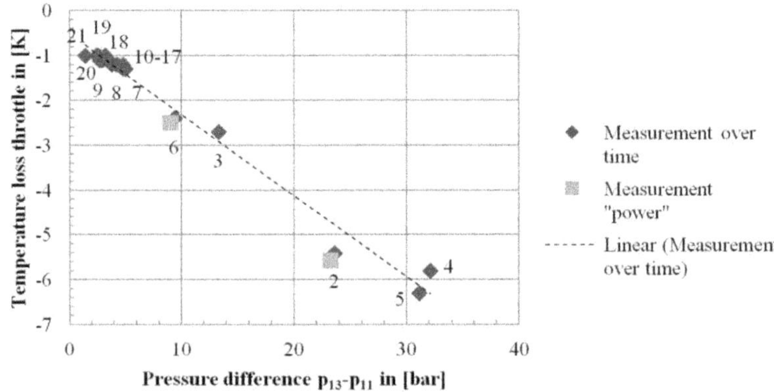

FIGURE A.5: Temperature loss of the throttle over the pressure drop of the throttle (Joule-Thompson)

TABLE A.2: Discharge "Power" (Date: 07.+08.10.2010)

	Unit		1	2	3
Air Mass Flow Rate	kg/s		199.92	300.78	402.44
Turbine Power	$MW_{(el)}$		105.28	207.83	318.91
Temperature	^oC	T_{11}	35.95	35.425	33.52
		T_{12}	27.405	29.86	31.03
		T_{13}	531.42	513.08	501.22
		T_{14}	309.7	310.17	309.82
		T_{15}	705.14	821.36	944.28
		T_{16}	390.23	430.43	473.85
Pressure	bar	p_{10}	53.565	52.065	50.4
		p_{11}	55.49	51.79	47.42
		p_{13}	18.83	28.62	38.48
		p_{14}	4.6	8.13	11.79
		p_{16}	1.01	1.01	1.01

TABLE A.3: Discharge over time (Date: 04.05.2011)

	1	2	3	4	5	6	7	8	9	10	11	12
	8:30	8:53	10:28	10:37	10:57	11:10	11:54	11:56	12:00	12:30	13:00	13:30
$\dot{m}$	0	408	396	218	217	397	394	394	391	370	350	333
P_T	0	316	312	120	121	313	313	313	310	288	266	246
T_{11}	25.4	36.9	28.9	28.9	28.6	27.3	23.8	23.7	23.4	21.6	19.9	18.6
T_{12}	21.3	31.5	26.2	23.1	22.3	24.9	22.5	22.4	22.2	20.4	18.7	17.4
T_{13}	45	500	496	520	521	499	497	497	498	503	506	507
T_{14}	65.6	262	259	297	300	259	257	256	257	257	261	265
T_{15}	na	938	937	724	725	938	938	938	936	909	886	864
T_{16}	18.3	469	471	397	398	472	471	472	471	462	454	446
p_{10}	65.5	63.9	53.7	53.2	52.3	51.2	47	46.9	46.5	44	41.6	39.4
p_{11}	na	61.6	51.2	52.9	51.9	47.4	43	42.8	42.5	40	38	35.9
p_{12}	na	38	37.9	20.8	20.8	38	38	38	37.7	35.7	33.8	32.1
p_{14}	na	11.7	11.6	5.1	5.2	11.6	11.6	11.6	11.5	10.8	10.1	9.4
p_{16}	1.015	1.015	1.014	1.014	1.015	1.016	1.015	1.016	1.016	1.015	1.015	1.015

	13	14	15	16	17	18	19	20	21	22	23	24
	14:00	14:30	15:00	15:30	16:00	16:30	16:45	16:48	16:49	16:54	16:55	17:00
$\dot{m}$	316	300	284	271	257	244	239	238	237	23	24	0
P_T	228	209	192	178	164	151	144	143	143	0	0	0
T_{11}	17.4	16.4	15.5	14.7	14	13.4	13.1	13	13	13.5	13.6	13.9
T_{12}	16.3	15.3	14.5	13.6	12.9	12.4	12.1	12	12	9.8	9.1	9.7
T_{13}	505	511	514	515	517	519	521	521	521	489	469	415
T_{14}	273	277	283	288	292	297	298	298	299	267	253	286
T_{15}	843	824	806	789	773	756	752	751	750	631	602	596
T_{16}	438	432	425	420	414	409	407	407	407	395	370	369
p_{10}	37.4	35.5	33.7	32	30.5	29	28.3	28.2	28.2	28.1	28.1	28.3
p_{11}	34	32.3	30.7	29.1	27.6	26.3	25.6	25.4	24.4	27.3	27.4	27.5
p_{12}	30.5	28.9	27.5	26.2	24.9	23.7	23.1	23	23	na	na	na
p_{14}	8.8	8.2	7.7	7.2	6.7	6.2	6	6	5.9	0.1	0.1	na
p_{16}	1.015	1.015	1.016	1.016	1.016	1.016	1.016	1.016	1.016	1.016	1.016	1.016

TABLE A.4: Huntorf operational data of a discharging trial in 2011

Time [h]	Mass flow rate $\dot{m}$ [kg/s]	$T_{wellhead}$ [°C]	$p_{wellhead}$ [bar]
08:30	0	25,4	65,5
08:53	-408	36,9	63,9
10:28	-396	28,9	53,7
10:37	-218	28,9	53,2
10:57	-217	28,6	52,3
11:10	-397	27,3	51,2
11:54	-394	23,8	47
11:56	-394	23,7	46,9
12:00	-391	23,4	46,5
12:30	-370	21,6	44
13:00	-350	19,9	41,6
13:30	-333	18,6	39,4
14:00	-316	17,4	37,4
14:30	-300	16,4	35,5
15:00	-284	15,5	33,7
15:30	-271	14,7	32
16:00	-257	14	30,5
16:30	-244	13,4	29
16:45	-239	13,1	28,3
16:48	-238	13	28,2
16:49	-237	13	28,2
16:54	-23	13,5	28,1
16:55	-24	13,6	28,1
17:00	0	13,9	28,3

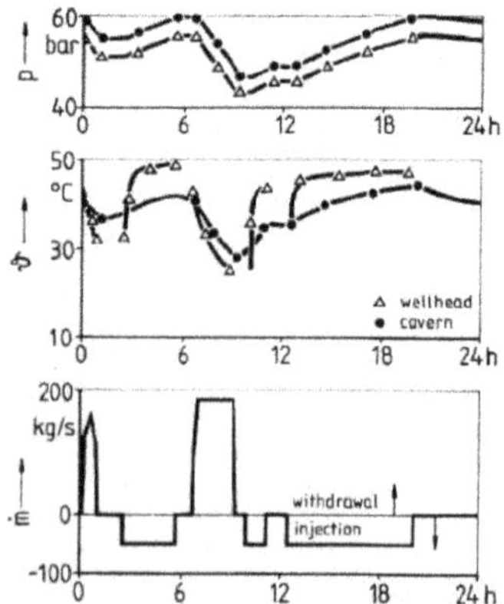

FIGURE A.6: Measured data from the Huntorf CAES commissioning published by Quast and Crotogino Quast and Crotogino, 1979: "Pressure, temperature, flow rate vs time (NK 1 observation)"

TABLE A.5: Huntorf operational data during commissioning from Quast and Crotogino Quast and Crotogino, 1979, originally published as diagram, see Fig. A.6.

Time	$\dot{m}$	Time	p_{cavern}	$p_{wellh.}$	Time	T_{cavern}	Time	$T_{wellh.}$
[h]	[kg/s]	[h]	[bar]	[bar]	[h]	[°C]	[h]	[°C]
00:00	0	00:00	59	55	00:00	45	00:00	45
00:15	150	01:00	55	51	00:30	40	00:30	37
00:30	170	03:00	56	52	01:00	37	01:00	32
00:45	150	05:30	60	56	02:30	38	02:30	32,5
01:00	0	06:45	59,8	55,8	02:45	39	02:45	42
02:15	0	08:00	54	49	04:00	41	04:00	48
02:30	-50	09:15	47	43,5	05:30	42	05:30	49
05:45	-50	11:30	49,5	46	06:30	41	06:30	43
06:00	0	12:50	49,5	46	07:45	34	07:15	33
06:30	0	14:45	53	49	09:00	28	08:45	26
06:45	180	17:10	57	53	10:00	30	10:00	36
09:00	180	19:50	60	56	11:00	35	11:00	44
09:15	0	00:00	59	55	12:30	35,5	13:00	46
09:45	0				14:25	40	15:20	46,5
10:00	-50				17:30	43	17:30	48
11:15	-50				20:00	44	19:30	48
11:30	0				23:59	40		
12:15	0							
12:30	-50							
20:00	-50							
20:15	0							

Bibliography

Alioth LLC. *Inflation calculator*. URL: http://www.in2013dollars.com/ (visited on 07/14/2017).

Allen, R. D., T. J. Doherty, and A. F. Fossum (1982). *Geotechnical issues and guidelines for storage of compressed air in excavated hard rock caverns: Report prepared for the U.S. Department of Energy by Pacific Northwest Laboratory*. Ed. by PNL. Richland, Washington. DOI: 10.2172/5437632.

Allen, R. D., T. J. Doherty, and R. L. Thoms (1982). *Geotechnical Factors and Guidlines for Compressed Air in Solution Mined Salt Cavities: Report prepared for the U.S. Department of Energy by Pacific Northwest Laboratory*. Ed. by PNL. Richland, Washington.

Alstom Power, Ecofys, E.ON Energie, KBB, IAEW, REpower, and Vattenfall Europe Transmission, eds. (2007). *Verbesserte Integration grosser Windstrommengen durch Zwischenspeicherung mittels CAES: Endbericht*.

Anderson, R., L. Bates, E. Johnson, and J. F. Morris (2015). "Packed bed thermal energy storage: A simplified experimentally validated model". In: *Journal of Energy Storage* 4, pp. 14–23. ISSN: 2352152X. DOI: 10.1016/j.est.2015.08.007.

Arabkoohsar, A., L. Machado, M. Farzaneh-Gord, and R. Koury (2015). "The first and second law analysis of a grid connected photovoltaic plant equipped with a compressed air energy storage unit". In: *Energy* 87, pp. 520–539. ISSN: 03605442. DOI: 10.1016/j.energy.2015.05.008.

Arsie, I., V. Marano, G. Nappi, and G. Rizzo (April 5-7, 2005). "A Model of a Hybrid Power Plant with Wind Turbines and Compressed Air Energy Storage". In: *Proceedings of PWR2005*. Ed. by ASME Power, pp. 1–14.

Ausfelder, F., C. Beilmann, M. Bertau, S. Bräuninger, A. Heinzel, R. Hoer, W. Koch, F. Mahlendorf, A. Metzelthin, M. Peuckert, L. Plass, K. Räuchle, M. Reuter, G. Schaub, S. Schiebahn, E. Schwab, F. Schüth, D. Stolten, G. Teßmer, K. Wagemann, and K.-F. Ziegahn (2015). "Energiespeicherung als Element einer sicheren Energieversorgung". In: *Chemie Ingenieur Technik* 87.1-2, pp. 17–89. ISSN: 0009286X. DOI: 10.1002/cite.201400183.

Baehr, H. D. and S. Kabelac (2009). *Thermodynamik: Grundlagen und technische Anwendungen*. 14. Aufl. Springer-Lehrbuch. Berlin, Heidelberg: Springer-Verlag Berlin Heidelberg. ISBN: 9783642005558.

Bailie, B. (2017-07-12). *Cost of CAES: phone call*. Ed. by F. Kaiser.

Barbour, E., D. Mignard, Y. Ding, and Y. Li (2015). "Adiabatic Compressed Air Energy Storage with packed bed thermal energy storage". In: *Applied Energy* 155, pp. 804–815. ISSN: 03062619. DOI: 10.1016/j.apenergy.2015.06.019.

Bauer, M. (2001). "Theoretische und experimentelle Untersuchungen zum Wärmetransport in gasdurchströmten Festbettrohrreaktoren". doctoral thesis. Halle: Martin-Luther-Universität Halle-Wittenberg.

Beasley, D. E. and J. A. Clark (1984). "Transient response of a packed bed for thermal energy storage". In: *Int. J. Heat Mass Transfer* 27.9, pp. 1659–1668.

Beck, H.-P., B. Engel, L. Hofmann, R. Menges, T. Turek, and H. Weyer (2013). *Eignung von Speichertechnologien zum Erhalt der Systemsicherheit: Abschlussbericht*. Vol. 13. Schriftenreihe des Energie-Forschungszentrums Niedersachsen. Göttingen: Cuvillier Verlag. ISBN: 978-3-95404-439-9.

Beckwith & Associates (1983). *Review of environmental studies and issues on compressed-air energy storage: Report prepared for the U.S. Department of Energy by Pacific Northwest Laboratory*. Ed. by PNL. Richland, Washington.

Berest, P. and B. Brouard (2003). "Safety of Salt Caverns Used for Underground Storage: Blow Out; Mechanical Instability; Seepage; Cavern Abandonment". In: *Oil Gas Science Technology* 58.3, pp. 361–384.

Berrada, A. and K. Loudiyi (2016). "Operation, sizing, and economic evaluation of storage for solar and wind power plants". In: *Renewable and Sustainable Energy Reviews* 59, pp. 1117–1129. ISSN: 13640321. DOI: 10.1016/j.rser.2016.01.048.

BGR InSpEE. *Geoviewer Bundesanstalt für Geowissenschaften und Rohstoffe, Arbeitsbereich Geodatenmanagement, InSpEE Webdienst Salzstrukturen*. URL: https://geoviewer.bgr.de/ (visited on 10/17/2016).

Biasi, V. de (2009). "New Solutions for energy storage and smart grid load management". In: *Gas Turbine World* 39.2, pp. 22–25.

BMWi, ed. (2014). *Abschlussbericht Metastudie Energiespeicher: Studie im Auftrag des Bundesministeriums für Wirtschaft und Energie (BMWi)*. Oberhausen and Kassel.

Bollinger, B. (2015). *Technology Performance Report SustainX Smart Grid Program: Demonstration of Promising Energy Storage Technologies, Project Report*. Ed. by I. SustainX. Seabrook.

Borgnakke, C. and R. C. Sonntag (2009). *Fundamentals of thermodynamics*. 7th ed., International student version, SI version. Hoboken, N.J: Wiley. ISBN: 978-0-470-17157-8.

Bosio, F. de and V. Verda (2015). "Thermoeconomic analysis of a Compressed Air Energy Storage (CAES) system integrated with a wind power plant in the framework of the IPEX Market". In: *Applied Energy* 152, pp. 173–182. ISSN: 03062619. DOI: 10.1016/j.apenergy.2015.01.052.

Böttcher, N., U.-J. Görke, O. Kolditz, and T. Nagel (2017). "Thermo-mechanical investigation of salt caverns for short-term hydrogen storage". In: *Environmental Earth Sciences* 76.3. ISSN: 1866-6280. DOI: 10.1007/s12665-017-6414-2.

Brendel, R. and R. Niepelt (2016). *Abschlussbericht zum Auftrag Zusatzgutachten Zeitlich höher aufgelöste Szenarien zum Hauptgutachten Szenarien zur Energieversorgung in Niedersachsen im Jahr 2050 der CUTEC GmbH an die ISFH GmbH*. Emmerthal.

Briola, S., P. Di Marco, R. Gabbrielli, and J. Riccardi (2016). "A novel mathematical model for the performance assessment of diabatic compressed air energy storage systems including the turbomachinery characteristic curves". In: *Applied Energy* 178, pp. 758–772. ISSN: 03062619. DOI: 10.1016/j.apenergy.2016.06.091.

Brown Boveri & Cie (1980). "Huntorf Air Storage Gas Turbine Power Plant". In: *Energy Supply* D GK 90 202 E, pp. 2–14.

— (1986). "Operating Experience with the Huntorf Air Storage Gas Turbine Power Station: D GK 1274 86 E". In: *Brown Boveri Review* 73.6, pp. 297–305.

Budt, M., D. Wolf, R. Span, and J. Yan (2016a). "A review on compressed air energy storage: Basic principles, past milestones and recent developments". In: *Applied Energy* 170, pp. 250–268. ISSN: 03062619. DOI: 10.1016/j.apenergy.2016.02.108.

— (2016b). "Compressed Air Energy Storage – An Option for Medium to Large Scale Electrical-energy Storage". In: *Energy Procedia* 88, pp. 698–702. ISSN: 18766102. DOI: 10.1016/j.egypro.2016.06.046.

Bullough, C., C. Gatzen, C. Jakiel, M. Koller, A. Nowi, and S. Zunft (2004). "Advanced Adiabatic Compressed Air Energy Storage for the Integration of Wind Energy". In: *Proceedings of the European Wind Energy Conference, EWEC.*

Bundesministerium für Wirtschaft und Technologie and Bundesministerium für Umwelt, Naturschutz und Reaktorsicherheit, eds. (2012). *Erster Monitoring-Bericht: Energie der Zukunft*. Berlin.

Bush, J. B., J. M. Campbell, P. M. Jarvis, D. A. Plionis, P. A. Aamodt, R. D. Anderson, J. L. Ash, A. S. Mitchell, L. A. Rives, G. A. Englesson, and E. J. Sosnowicz (1976). *Economic and Technical Feasability Study Of Compressed Air Storage: Final Report prepared for U.S. Energy Research and Development Administration, Washington, D.C.* Ed. by General Electric. Washington, D.C.

Carnot, S. (1825). *Reflections on the motive power of fire, and on machines fitted to develop that power: excerpts.*

Cascetta, M., G. Cau, P. Puddu, and F. Serra (2014). "Numerical Investigation of a Packed Bed Thermal Energy Storage System with Different Heat Transfer Fluids". In: *Energy Procedia* 45, pp. 598–607. ISSN: 18766102. DOI: 10.1016/j.egypro.2014.01.064.

— (2015). "A Study of a Packed-bed Thermal Energy Storage Device: Test Rig, Experimental and Numerical Results". In: *Energy Procedia* 81, pp. 987–994. ISSN: 18766102. DOI: 10.1016/j.egypro.2015.12.157.

— (2016). "A comparison between CFD simulation and experimental investigation of a packed-bed thermal energy storage system". In: *Applied Thermal Engineering* 98, pp. 1263–1272. ISSN: 13594311. DOI: 10.1016/j.applthermaleng.2016.01.019.

Castellani, B., A. Presciutti, M. Filipponi, A. Nicolini, and F. Rossi (2015). "Experimental Investigation on the Effect of Phase Change Materials on Compressed Air Expansion in CAES Plants". In: *Sustainability* 7.8, pp. 9773–9786. ISSN: 2071-1050. DOI: 10.3390/su7089773.

Chen, H., T. N. Cong, W. Yang, C. Tan, Y. Li, and Y. Ding (2009). "Progress in electrical energy storage system: A critical review". In: *Progress in Natural Science* 19.3, pp. 291–312. ISSN: 10020071. DOI: 10.1016/j.pnsc.2008.07.014.

Chiesa, P., G. Lozza, and L. Mazzocchi (2005). "Using Hydrogen as Gas Turbine Fuel". In: *Journal of Engineering for Gas Turbines and Power* 127.1, p. 73. ISSN: 07424795. DOI: 10.1115/1.1787513.

Crotogino, F. (2003). "Druckluftspeicher-GT-Kraftwerke: Ausgleich fluktuierender Stromproduktion". In: *etz elektrotechnik & automation* 5, pp. 12–18.

Crotogino, F. and R. Hamelmann (2007). "Wasserstoff-Speicherung in Salzkavernen zur Glättung des Windstromangebots". In: *Nutzung regenerativer Energiequellen und Wasserstofftechnik 2007*. Ed. by T. Luschtinetz and J. Lehmann. Stralsund: Fachhochschule Stralsund. ISBN: 3-9809953-6-4. URL: http://www.fh-stralsund.de/ps/tools/download.php?file=/internet/dms/psfile/docfile/45/Tagungsban53d8ef5b11bc8.pdf&name=Tagungsband_2007.pdf&disposition=inline (visited on 03/01/2017).

Crotogino, F., K.-U. Mohmeyer, and R. Scharf (2001-04-15). *Huntorf CAES: More than 20 years of successful operation: Report Spring 2001 Meeting, Orlando*. Orlando, Florida, USA.

Crotogino, F., S. Donadei, U. Bünger, and H. Landinger (2010). "Large Scale Hydrogen Underground Storage for Securing Future Energy Supply". In: *18th World Hydrogen Energy Conference 2010 - WHEC 2010*. Ed. by D. Stolten and T. Grube. Schriften des Forschungszentrums Jülich Reihe Energie & Umwelt. Jülich: Forschungszentrum IEF-3, pp. 36–45. ISBN: 9783893366545.

Dehmer, D. (2013). "The German Energiewende: The First Year". In: *The Electricity Journal* 26.1, pp. 71–78. ISSN: 10406190. DOI: 10.1016/j.tej.2012.12.001.

Dietz, P., ed. (2008). *Grundlast von der Nordsee: Abschlussbericht in Rahmen der Projektstudie Netzintegration von Offshore-Grosswindanlagen*. Clausthal-Zellerfeld: Papierflieger. ISBN: 978-3-89720-978-7.

Doering, E., H. Schedwill, and M. Dehli (2008). *Grundlagen der technischen Thermodynamik: Lehrbuch für Studierende der Ingenieurwissenschaften ; mit 45 Tabellen sowie 56 Aufgaben mit Lösungen*. 6., überarb. und erw. Aufl. Studium. Wiesbaden: Vieweg + Teubner. ISBN: 9783835101494. URL: http://deposit.d-nb.de/cgi-bin/dokserv?id=2997828&prov=M&dok_var=1&dok_ext=htm.

Donadei, S. and D. Zander-Schiebenhöfer (22./23. April 2015). *Bestimmung des Speicherpotenzials Erneuerbarer Energien in den Salzstrukturen Norddeutschlands: Vorstellung des Projektes InSpEE*. Celle.

Donadei, S., D. Zander-Schiebenhöfer, P. L. Horvath, D. Zapf, K. Staudtmeister, R. B. Rokahr, S. Fleig, L. Pollok, M. Hölzner, J. Hammer, S. Gast, C. Riesenberg, and G. von Goerne (2015). "Project InSpEE - Rock Mechanical Design for CAES and H2 Storage Caverns & Evaluation of Storage Capacity in NW-Germany". In: *The Third Sustainable Earth Sciences Conference and Exhibition*. Proceedings. EAGE Publications BVNetherlands. DOI: 10.3997/2214-4609.201414255.

Dötsch, C., A. Kanngießer, and D. Wolf (2009). "Speicherung elektrischer Energie – Technologien zur Netzintegration erneuerbarer Energien". In: *uwf UmweltWirtschaftsForum* 17.4, pp. 351–360. ISSN: 0943-3481. DOI: 10.1007/s00550-009-0150-3.

Dresen (2010). "Advanced Systematic Determination of the Capacities of Salt Caverns for Undergrounde Gas Storage". In: *Proc. SMRI Fall Technical Conference 2010 Leipzig*.

Dresser-Rand Group Inc. (2013). *Dresser-Rand Awarded Compressed Air Energy Storage (CAES) Project: Total Project Value Approximately $200 Million*. Houston. URL: http://www.prnewswire.com/news-releases/dresser-rand-awarded-compressed-air-energy-storage-caes-project-213826301.html (visited on 07/14/2017).

Drost, M. K., F. R. Zaloudek, and W. V. Loscutoff (1980). *A Preliminary Study of A-CAES in Aquifers: Report prepared for the U.S. Department of Energy by Pacific Northwest Laboratory*. Ed. by PNL. Richland, Washington.

Drury, E., P. Denholm, and R. Sioshansi (2011). "The value of compressed air energy storage in energy and reserve markets". In: *Energy* 36.8, pp. 4959–4973. ISSN: 03605442. DOI: 10.1016/j.energy.2011.05.041.

DVGW Deutsche Vereinigung des Gas- und Wasserfaches e. V. (Mai 2008). *Arbeitsblatt G 260: Gasbeschaffenheit*. Bonn.

Ebert, A. (1992). "Analyse und Bewertung verschiedener Pumpgrenzkriterien für vielstufige Verdichter". Diplomarbeit. Dresden: Technische Universität Dresden.

Elmegaard, B. and W. Brix (2011). *Efficiency of Compressed Air Energy Storage: Report, Technical University of Denmark, Department of Mechanical Engineering*. Technical University of Denmark.

EPRI and the U.S. Department of Energy, ed. (2003). *EPRI-DOE Handbook of Energy Storage for Transmission and Distribution Applications: Technical Report*. Palo Alto, California.

Erikson, R. L. (1983). *Thermophysical Behavior of StPeter Sandstone Application to CAES in an Aquifer: Report prepared for the U.S. Department of Energy by Pacific Northwest Laboratory*. Ed. by PNL. Richland, Washington.

EWE-Netz GmbH (2015). *Analysebericht Erdgas: GB, Erdgasspeicher Huntorf*.

Faulstich, M., H.-P. Beck, H.-H. Schmidt-Kanefendt, F. Kaiser, C. Yilmaz, J. Zum Hingst, and J. Kuck (2016a). *Szenarien zur Energieversogung in Niedersachsen im Jahr 2050: Zusatzgutachten zeitlich höher aufgelöste Szenarien*. Hannover.

Faulstich, M., H.-P. Beck, F. Kaiser, J. Kuck, H.-H. Schmidt-Kanefendt, G. Römer, C. Yilmaz, W. Siemers, and S. Ronia (2016b). *Szenarien zur Energieversorgung in Niedersachsen im Jahr 2050: Gutachten: Abschlussbericht zum Angebot Nr. 703500*. Hannover. ISBN: 978-3-00-052763-0.

Fertig, E. and J. Apt (2011). "Economics of compressed air energy storage to integrate wind power: A case study in ERCOT". In: *Energy Policy* 39.5, pp. 2330–2342. ISSN: 03014215. DOI: 10.1016/j.enpol.2011.01.049.

Fichtner GmbH, ed. (2014). *Erstellung eines Entwicklungskonzeptes Energiespeicher in Niedersachsen: Studie Juli 2014*. Stuttgart.

Foley, A. and I. Díaz Lobera (2013). "Impacts of compressed air energy storage plant on an electricity market with a large renewable energy portfolio". In: *Energy* 57, pp. 85–94. ISSN: 03605442. DOI: 10.1016/j.energy.2013.04.031.

Fort, J. A. (1982). *CAESCAP: A Computer Code for Compressed Air Energy Storage Plant Cycle Analysis: Report prepared for the U.S. Department of Energy by Pacific Northwest Laboratory*. Ed. by PNL.

— (1983). *Thermodynamic analysis of five compressed-air energy-storage cycles: Report prepared for the U.S. Department of Energy by Pacific Northwest Laboratory*. Ed. by PNL.

Fraunhofer Institute for Solar Energy Systems ISE. *Energy Charts: Preise*. Ed. by B. Burger. URL: https://www.energy-charts.de/price_de.htm (visited on 03/07/2017).

Gaggioli, R. A., D. H. Richardson, and A. J. Bowman (2002). "Available Energy—Part I: Gibbs Revisited". In: *Journal of Energy Resources Technology* 124.2, p. 105. ISSN: 01950738. DOI: 10.1115/1.1448336.

Garvey, S. D. (2012). "The dynamics of integrated compressed air renewable energy systems". In: *Renewable Energy* 39.1, pp. 271–292. ISSN: 09601481. DOI: 10.1016/j.renene.2011.08.019.

— (2015). *Two Novel Configurations for Compressed Air Energy Storage Exploiting High-Grade Thermal Energy Storage: presented at UK-China Thermal Energy Storage Forum, Beijing*. Beijing.

Gatzen, C. (2008). *The economics of power storage: Theory and empirical analysis for Central Europe*. Vol. Bd. 63. Schriften des energiewirtschaftlichen Instituts. München: Oldenbourg Industrieverlag. ISBN: 978-3-835631380.

Gay, F. W. (1948). *Means for storing fluids for power generation: US Patent 2433896*.

Giramonti, A. and R. D. Lessard (1974). "Exploratory Evaluation of Compressed Air Storage Peak-Power Systems". In: *Energy Sources* 1.3, pp. 283–294. ISSN: 0090-8312. DOI: 10.1080/00908317408945926.

Gobbato, P., M. Masi, A. Toffolo, and A. Lazzaretto (2011). "Numerical simulation of a hydrogen fuelled gas turbine combustor". In: *International Journal of Hydrogen Energy* 36.13, pp. 7993–8002. ISSN: 03603199. DOI: 10.1016/j.ijhydene.2011.01.045.

Goodson, J. O. (1993). *History of First U.S. Compressed Air Energy Storage (CAES) Plant (110-MW-26 h): Volume 1: Early CAES Development*. Ed. by Electric Power Research Insitute.

Grazzini, G. and A. Milazzo (2008). "Thermodynamic analysis of CAES / TES systems for renewable energy plants". In: *Renewable Energy* 33.9, pp. 1998–2006. ISSN: 09601481. DOI: 10.1016/j.renene.2007.12.003.

Greenblatt, J. B., S. Succar, D. C. Denkenberger, R. H. Williams, and R. H. Socolow (2007). "Baseload wind energy: modeling the competition between gas turbines and compressed air energy storage for supplemental generation". In: *Energy Policy* 35.3, pp. 1474–1492. ISSN: 03014215. DOI: 10.1016/j.enpol.2006.03.023.

Gu, Y., J. McCalley, M. Ni, and R. Bo (2013). "Economic Modeling of Compressed Air Energy Storage". In: *Energies* 6.4, pp. 2221–2241. ISSN: 1996-1073. DOI: 10.3390/en6042221.

Hagoort, J. (1994). "Simulation of Production and Injection Performance of Gas Storage Caverns in Salt Formations". In: *SPE Reservoir Engineering*, pp. 278–282.

Hartmann, N., O. Vöhringer, C. Kruck, and L. Eltrop (2012a). "Simulation and analysis of different adiabatic Compressed Air Energy Storage plant configurations". In: *Applied Energy* 93, pp. 541–548. ISSN: 03062619. DOI: 10.1016/j.apenergy.2011.12.007.

Hartmann, N., L. Eltrop, N. Bauer, J. Salzer, S. Schwarz, and M. Schmidt (2012b). *Stromspeicherpotenziale für Deutschland: Report by zfes, Universität Stuttgart*. Ed. by zfes. Universität Stuttgart.

Hendrickson, P. L. (1981). *Legal and Regulatory issues affecting Compressed Air Energy Storage: Report prepared for the U.S. Department of Energy by Pacific Northwest Laboratory*. Ed. by PNL. Richland, Washington.

Hobson, M. J., R. Kohanski, S. Bahadur, C. L. Driggs, D. C. James, J. C. Sangermano, J. C. Gerhardt, R. C. Hinkel, K. A. Hresko, M. K. Galola, D. Birkholz, R. Vincent, G. Hughes, H. Miller, L. McGuire, R. Williams, P. E. Chew, J. L. Nash-Webber, W. V. Loscutoff, L. D. Kannberg, and F. R. Zaloudek (1981). *Conceptual design and engineering studies of Adiabatic CAES with Thermal Energy Storage: Report prepared for the U.S. Department of Energy by Pacific Northwest Laboratory*. Ed. by PNL. Richland, Washington.

Hoffeins, H., D. Romeyke, and F. Sütterlin (1980). "Die Inbetriebnahme der ersten Luftspeicher-Gasturbinengruppe: Energieversorgung: Druckschrift Nr. CH GK 1139 81 D". In: *Brown Boveri Mitteilungen* 67.8, pp. 465–473.

Horvath, P. L., S. Donadei, and Schneider (2016). *GIS-basierte Potenzialabschätzung für Salzkavernen und Aufbau des Informationssystems Salz im Projekt InSpEE: Poster*. Ed. by KBB UT. Celle.

Hostetler, D. D., S. W. Childs, and S. J. Phillips (1983). *Subsurface Monitoring of Reservoir Pressure, Temperature, Relative humidity, and Water Content at Pitsfield, Illinois: System Design: Report prepared for the U.S. Department of Energy by Pacific Northwest Laboratory*. Ed. by PNL. Richland, Washington.

HyUnder. URL: http://www.hyunder.eu (visited on 09/01/2016).

Ibrahim, H., R. Younes, A. Ilinca, and J. Perron (2007). "Investigation des générateurs hybrides d'électricité de type éolien-air comprimé". In: *Revue des Energies Renouvelables*, pp. 47–50.

Ibrahim, H., A. Ilinca, D. Rousse, Y. Dutil, and J. Perron (2012). *Analyse des systèmes de génération d'électricité pour les sites isolés basés sur l'utilisation du stockage d'air comprimé en hybridation avec un jumelage éolien-diesel*.

Jakiel, C., S. Zunft, and A. Nowi (2007). "2007_Jakiel_Adiabatic compressed air energy storage plants for efficient peak load_AA-CAES". In:

Juste, G. L. (2006). "Hydrogen injection as additional fuel in gas turbine combustor: Evaluation of effects". In: *International Journal of Hydrogen Energy* 31.14, pp. 2112–2121. ISSN: 03603199. DOI: 10.1016/j.ijhydene.2006.02.006.

Kaiser, F. (2012-06-28). "Potentiale der Herstellung synthetischer Kohlenwasserstoffe zur Speicherung regenerativen Stroms". Diplomarbeit. Clausthal-Zellerfeld: Technische Universität Clausthal.

— (2016). *Economic equivalent energy storage efficiency for Compressed Air Energy Storage (CAES)*. DOI: 10.13140/RG.2.1.4989.9125.

Kaiser, F. and W. Busch (2015). "Der beste Stromspeicher? - Pumpspeicher und die Alternativen". In: *Pumpspeicher für die Energiewende - Spitzentechnologie auf Eis?* Ed. by W. Busch and F. Kaiser. Schriftenreihe des Energie-Forschungszentrums Niedersachsen. Göttingen: Cuvillier Verlag, pp. 72–87. ISBN: 978-3-7369-9130-9. DOI: 10.13140/RG.2.1.4341.4803.

Kaiser, F. and U. Krüger (2019). "Exergy analysis and assessment of performance criteria for compressed air energy storage concepts". In: *International Journal of Exergy* 28.3, pp. 229–254. DOI: 10.1504/IJEX.2019.10020032.

Kaiser, F., R. Weber, and U. Krüger (2018). "Thermodynamic Steady-State Analysis and Comparison of Compressed Air Energy Storage (CAES) Concepts". In: *International Journal of Thermodynamics* 21.3, pp. 144–156. ISSN: 1301-9724. DOI: 10.5541/ijot.407824.

Kaldellis, J. K. and D. Zafirakis (2007). "Optimum energy storage techniques for the improvement of renewable energy sources-based electricity generation economic efficiency". In: *Energy* 32.12, pp. 2295–2305. ISSN: 03605442. DOI: 10.1016/j.energy.2007.07.009.

Kaldemeyer, C., C. Boysen, and I. Tuschy (2016). "Compressed Air Energy Storage in the German Energy System – Status Quo & Perspectives". In: *Energy Procedia* 99, pp. 298–313. ISSN: 18766102. DOI: 10.1016/j.egypro.2016.10.120.

Kannberg, L. D. (1981). *Advanced Concepts Second Generation CAES: Report prepared for the U.S. Department of Energy by Pacific Northwest Laboratory*. Ed. by PNL. Richland, Washington.

— (1983). *Underground Energy Storage Program_Annual Report: Report prepared for the U.S. Department of Energy by Pacific Northwest Laboratory*. Ed. by PNL. Richland, Washington.

Khaitan, S. K. and M. Raju (2012). "Dynamics of hydrogen powered CAES based gas turbine plant using sodium alanate storage system". In: *International Journal of Hydrogen Energy* 37.24, pp. 18904–18914. ISSN: 03603199. DOI: 10.1016/j.ijhydene.2012.10.004.

— (2013). "Dynamic simulation of air storage-based gas turbine plants". In: *International Journal of Energy Research* 37.6, pp. 558–569. ISSN: 0363907X. DOI: 10.1002/er.1944.

Khaledi, K., E. Mahmoudi, M. Datcheva, and T. Schanz (2016). "Analysis of compressed air storage caverns in rock salt considering thermo-mechanical cyclic loading". In: *Environmental Earth Sciences* 75.15. ISSN: 1866-6280. DOI: 10.1007/s12665-016-5970-1.

Kim, Y.-M., J.-H. Lee, S.-J. Kim, and D. Favrat (2012). "Potential and Evolution of Compressed Air Energy Storage: Energy and Exergy Analyses". In: *Entropy* 14.12, pp. 1501–1521. ISSN: 1099-4300. DOI: 10.3390/e14081501.

Klumpp, F. (2016). "Comparison of pumped hydro, hydrogen storage and compressed air energy storage for integrating high shares of renewable energies—Potential, cost-comparison and ranking". In: *Journal of Energy Storage* 8, pp. 119–128. ISSN: 2352152X. DOI: 10.1016/j.est.2016.09.012.

Kondoh, J., I. Ishii, H. Yamaguchi, A. Murata, K. Otani, K. Sakuta, N. Higuchi, S. Sekine, and M. Kamimoto (2000). "Electrical energy storage systems for energy networks". In: *Energy*

Conversion and Management 41.17, pp. 1863–1874. ISSN: 01968904. DOI: 10.1016/S0196-8904(00)00028-5.

Kreid, D. K. (1976). *Technical and economic feasibility analysis of the no-fuel compressed air energy storage concept*. Richland, Washington. DOI: 10.2172/7153562.

Kreid, D. (1977). *Analysis of Advanced Compressed Air Energy Storage Concepts*. Richland, Washington.

Kreid, D. K. and M. A. McKinnon (1978). *FY 1977 Progress report, Compressed air energy storage advanced systems analysis*. Richland, Washington. DOI: 10.2172/5010148.

Kruck, O., D. Zander-Schiebenhöfer, and J. Johansen (2013). "Thermodynamische Modellierung und Simulation im Zuge der Flutung und Rekomplettierung des Kavernenspeichers Ll. Torup To-8 in Dänemark". In: *Erdoel Erdgas Kohle* 129.11, pp. 1–6.

Krüger, O. (2015-02-19). "On the Influence of Steam in Premixed Hydrogen Flames for Future Gas Turbine Applications". doctoral thesis. Berlin: Technische Universität Berlin.

Krüger, U. (27.07.2015). *Daten aus dem Betrieb des Druckluftspeicherkraftwerks Huntorf: Datentabelle*. Ed. by F. Kaiser.

Kushnir, R., A. Dayan, and A. Ullmann (2012). "Temperature and pressure variations within compressed air energy storage caverns". In: *International Journal of Heat and Mass Transfer* 55.21-22, pp. 5616–5630. ISSN: 00179310. DOI: 10.1016/j.ijheatmasstransfer.2012.05.055.

Kushnir, R., A. Ullmann, and A. Dayan (2012a). "Thermodynamic and hydrodynamic response of compressed air energy storage reservoirs: a review". In: *Reviews in Chemical Engineering* 28.2-3, pp. 123–148. ISSN: 2191-0235. DOI: 10.1515/revce-2012-0006.

— (2012b). "Thermodynamic Models for the Temperature and Pressure Variations Within Adiabatic Caverns of Compressed Air Energy Storage Plants". In: *Journal of Energy Resources Technology* 134, pp. 021901–1 –021901–10. URL: http://energyresources.asmedigitalcollection.asme.org/ (visited on 01/27/2015).

Laing, D., T. Bauer, N. Breidenbach, B. Hachmann, and M. Johnson (2013). "Development of high temperature phase-change-material storages". In: *Applied Energy* 109, pp. 497–504. ISSN: 03062619. DOI: 10.1016/j.apenergy.2012.11.063.

Langham, E. J. (1965). "The underground storage of compressed air for gas turbines: a dynamic study on analoque computer". In: pp. 216–224.

LBEG (2012). *Erdöl und Erdgas in der Bundesrepublik Deutschland 2012: Landesamt für Bergbau und Geologie*. Hannover.

Leachman, J. W., R. T. Jacobsen, S. G. Penoncello, and E. W. Lemmon (2009). "Fundamental Equations of State for Parahydrogen, Normal Hydrogen, and Orthohydrogen". In: *Journal of Physical and Chemical Reference Data* 38.3, pp. 721–748. ISSN: 0047-2689. DOI: 10.1063/1.3160306.

Lechner, C. and J. Seume (2010). *Stationäre Gasturbinen*. Berlin, Heidelberg: Springer. ISBN: 978-3-540-92787-7. DOI: 10.1007/978-3-540-92788-4.

Lee, M. C., S. B. Seo, J. H. Chung, S. M. Kim, Y. J. Joo, and D. H. Ahn (2010). "Gas turbine combustion performance test of hydrogen and carbon monoxide synthetic gas". In: *Fuel* 89.7, pp. 1485–1491. ISSN: 00162361. DOI: 10.1016/j.fuel.2009.10.004.

Leithner, R. (20.01.2015). *Druckluftspeicherkraftwerke: Vortrag am Institut für Kraft-werkstechnik und Wärmeübertragung an der Leibnitz Universität Hannover*. Hannover.

Lemmon, E. W., R. T. Jacobsen, S. G. Penoncello, and D. G. Friend (2000). "Thermodynamic Properties of Air and Mixtures of Nitrogen, Argon, and Oxygen From 60 to 2000 K at Pressures to 2000 MPa". In: *J. Phys. Chem. Ref. Data* 29.3, pp. 331–385.

Leuger, B. and T. Beutel (2012). "Bewertung der Betriebserfahrungen mit der Gasspeicherkaverne Huntorf K6: Felsmechanik/Energiewende". In: *mining + geo* 4, pp. 674–677.

Liu, J.-L. and J.-H. Wang (2016). "A comparative research of two adiabatic compressed air energy storage systems". In: *Energy Conversion and Management* 108, pp. 566–578. ISSN: 01968904. DOI: 10.1016/j.enconman.2015.11.049.

Liu, Y., C.-K. Woo, and J. Zarnikau (2017). "Wind generation's effect on the ex post variable profit of compressed air energy storage: Evidence from Texas". In: *Journal of Energy Storage* 9, pp. 25–39. ISSN: 2352152X. DOI: 10.1016/j.est.2016.11.004.

Lund, H. and G. Salgi (2009). "The role of compressed air energy storage (CAES) in future sustainable energy systems". In: *Energy Conversion and Management* 50.5, pp. 1172–1179. ISSN: 01968904. DOI: 10.1016/j.enconman.2009.01.032.

Lund, P. D., J. Lindgren, J. Mikkola, and J. Salpakari (2015). "Review of energy system flexibility measures to enable high levels of variable renewable electricity". In: *Renewable and Sustainable Energy Reviews* 45, pp. 785–807. ISSN: 13640321. DOI: 10.1016/j.rser.2015.01.057.

Lux, K.-H. (2009). "Design of salt caverns for the storage of natural gas, crude oil and compressed air: Geomechanical aspects of construction, operation and abandonment". In: *Geological Society, London, Special Publications* 313.1, pp. 93–128. ISSN: 0305-8719. DOI: 10.1144/SP313.7.

Madlener, R. and J. Latz (2013). "Economics of centralized and decentralized compressed air energy storage for enhanced grid integration of wind power". In: *Applied Energy* 101, pp. 299–309. ISSN: 03062619. DOI: 10.1016/j.apenergy.2011.09.033.

Malyshenko, S. P., V. I. Prigozhin, A. R. Savich, A. I. Schastlivtsev, V. A. Il'ichev, and O. V. Nazarova (2012). "Effectiveness of steam generation in oxyhydrogen steam generators of the megawatt power class". In: *High Temperature* 50.6, pp. 765–773. ISSN: 0018-151X. DOI: 10.1134/S0018151X12050112.

Marano, V., G. Rizzo, and F. A. Tiano (2012). "Application of dynamic programming to the optimal management of a hybrid power plant with wind turbines, photovoltaic panels and compressed air energy storage". In: *Applied Energy* 97, pp. 849–859. ISSN: 03062619. DOI: 10.1016/j.apenergy.2011.12.086.

Martin, R. (2013-08-19). *Compressed Air Energy Storage to Experience Dramatic Growth over the Next 10 Years*. URL: http://www.navigantresearch.com/newsroom/compressed-air-energy-storage-to-experience-dramatic-growth-over-the-next-10-years (visited on 02/15/2017).

Marx, P. (2012). "Brennstoff- und Betriebsflexibilität anhand von Betriebserfahrungen mit Alstom Gasturbinenneuanlagen". In: *Kraftwerkstechnik*. Ed. by M. Beckmann and A. Hurtado. Neuruppin: TK Verlag, pp. 681–699. ISBN: 978-3-935317-87-0.

Mason, J. E. and C. L. Archer (2012). "Baseload electricity from wind via compressed air energy storage (CAES)". In: *Renewable and Sustainable Energy Reviews* 16.2, pp. 1099–1109. ISSN: 13640321. DOI: 10.1016/j.rser.2011.11.009.

Maton, J.-P., L. Zhao, and J. Brouwer (2013). "Dynamic modeling of compressed gas energy storage to complement renewable wind power intermittency". In: *International Journal of Hydrogen Energy* 38.19, pp. 7867–7880. ISSN: 03603199. DOI: 10.1016/j.ijhydene.2013.04.030.

Mauch, B., P. M. Carvalho, and J. Apt (2012). "Can a wind farm with CAES survive in the day-ahead market?" In: *Energy Policy* 48, pp. 584–593. ISSN: 03014215. DOI: 10.1016/j.enpol.2012.05.061.

Mazloum, Y., H. Sayah, and M. Nemer (2016). "Static and Dynamic Modeling Comparison of an Adiabatic Compressed Air Energy Storage System". In: *J. Energy Resour. Technol.* 138.6, p. 062001. DOI: 10.1115/1.4033399.

McBride, T., A. Bell, and D. Kepshire (2013). *ICAES Innovation Foam-Based Heat Exchange: SustainX, Inc. Report*. Ed. by I. SustainX. Seabrook.

Mei, S., J. Wang, F. Tian, L. Chen, X. Xue, Q. Lu, Y. Zhou, and X. Zhou (2015). "Design and engineering implementation of non supplementary fired compressed air energy storage system: TICC-500". In: *Science China Technological Sciences* 58.4, pp. 600–611. ISSN: 1674-7321. DOI: 10.1007/s11431-015-5789-0.

Meier, A., C. Winkler, and D. Wuillemin (1991). "Experiment for modelling high temperature rock bed storage". In: *Solar Energy Materials* 24.1-4, pp. 255–264. ISSN: 01651633. DOI: 10.1016/0165-1633(91)90066-T.

Mohamadabadi, H. S. (2014). "Techno-economic assessment of the need for bulk energy storage in low-carbon electricity systems with a focus on compressed air storage (CAES)". doctoral thesis.

Moser, P., R. Marquardt, S. Freund, M. Klafki, and P. Pijanovic (2012). "Adiabater Druckluftspeicher für die Elektrizitätsversorgung – Der nächste Schritt: ADELE-ING". In: *Kraftwerkstechnik*. Ed. by M. Beckmann and A. Hurtado. Vol. 4. Neuruppin: TK Verlag, pp. 803–813. ISBN: 978-3-935317-87-0.

Nakhamkin, M., E. Swensen, R. B. Schainker, and B. Mehta (1989). *CAES Plant Performance and Economics as a Function of Underground Salt Dome Storage Transient Processes: ASME Paper, Toronto*. Ed. by ASME. Toronto. URL: https://asmedigitalcollection.asme.org/.

Narbel, P. A. and J. P. Hansen (2014). "Estimating the cost of future global energy supply". In: *Renewable and Sustainable Energy Reviews* 34, pp. 91–97. ISSN: 13640321. DOI: 10.1016/j.rser.2014.03.011.

Nieland, J. D. (2004). *Salt Cavern Thermal Simulator Version 2.0 User's Manual: Topical Report RSI-1760*. Ed. by RESPEC. Rapid City, South Dakota.

Nielsen, L. (2013). *GuD-Druckluftspeicherkraftwerk mit Wärmespeicher: Promotion an der Technischen Universität Braunschweig*. Vol. 14. Schriftenreihe des Energie-Forschungszentrums Niedersachsen (EFZN). Göttingen: Cuvillier Verlag. ISBN: 978-3-95404-488-7.

Nielsen, L., D. Qi, N. Brinkmeier, and R. Leithner (22.-23.03.2012). *Druckluftspeicherkraftwerk zur Netzintegration erneuerbarer Energien ISACOAST-CC: Presentation: Dezentralisierung und Netzausbau*. Göttingen.

Oldhafer, N., M. Düren, S. Stenger, and R. Leithner (2014). "Druckluftspeicherkraftwerk als Kurzzeitspeicher, Langzeitspeicher und Schattenkraftwerk". In: *Kraftwerkstechnik 2014*. Ed. by M. Beckmann and A. Hurtado. Freiberg: SAXONIA, pp. 1–10. ISBN: 978-3-934409-62-0.

Osterle, J. F. (1991). "The Thermodynamics of Compressed Air Energy Storage". In: *Journal of Energy Resources Technology* 113.1, pp. 7–11.

Pape, C., N. Gerhardt, P. Härte, A. Scholz, R. Schwinn, T. Drees, A. Maaz, and .... (2014). *Speicherbedarf für Erneuerbate Energien - Speicheralternativen - Speicheranreiz - Überwindung rechtlicher Hemnisse: Endbericht*. Kassel.

Park, J.-W., D. Park, D.-W. Ryu, B.-H. Choi, and E.-S. Park (2014). "Analysis on heat transfer and heat loss characteristics of rock cavern thermal energy storage". In: *Engineering Geology* 181, pp. 142–156. ISSN: 00137952. DOI: 10.1016/j.enggeo.2014.07.006.

Park, J.-W., J. Rutqvist, D. Ryu, E.-S. Park, and J.-H. Synn (2016). "Coupled thermal-hydrological-mechanical behavior of rock mass surrounding a high-temperature thermal energy storage cavern at shallow depth". In: *International Journal of Rock Mechanics and Mining Sciences* 83, pp. 149–161. ISSN: 13651609. DOI: 10.1016/j.ijrmms.2016.01.007.

Pellow, M. A., C. J. M. Emmott, C. J. Barnhart, and S. M. Benson (2015). "Hydrogen or batteries for grid storage? A net energy analysis". In: *Energy Environ. Sci.* 8.7, pp. 1938–1952. ISSN: 1754-5692. DOI: 10.1039/C4EE04041D.

Pickard, W. F., N. J. Hansing, and A. Q. Shen (2009). "Can large-scale advanced-adiabatic compressed air energy storage be justified economically in an age of sustainable energy?" In: *Journal of Renewable and Sustainable Energy* 1.3, p. 033102. ISSN: 19417012. DOI: 10.1063/1.3139449.

Pimm, A. J., S. D. Garvey, and R. J. Drew (2011). "Shape and cost analysis of pressurized fabric structures for subsea compressed air energy storage". In: *Proceedings of the Institution of Mechanical Engineers, Part C: Journal of Mechanical Engineering Science* 225.5, pp. 1027–1043. ISSN: 0954-4062. DOI: 10.1177/0954406211399506.

Pollak, R. (1994). *History of First US Compressed-Air Energy Storage (CAES) Plant (110 MW, 26h): Volume 2: Construction: Final Report Electric Power Research Insitute (EPRI)*. Ed. by Electric Power Research Insitute. Palo Alto, California.

Pollok, L., S. Gast, M. Hölzner, S. Fleig, C. Riesenberg, J. Hammer, G. von Goerne, S. Donadei, P. L. Horvath, D. Zander-Schiebenhöfer, D. Zapf, K. Staudtmeister, and R. B. Rokahr (2015). "Project InSpEE, Storage Potential for Renewable Energies - Insights into Northern Germany's Salt Structure Inventory". In: *The Third Sustainable Earth Sciences Conference and Exhibition*. Proceedings. EAGE Publications BVNetherlands. DOI: 10.3997/2214-4609.201414256.

Qi, D. (2012). "Optimierung von Hochtemperaturwärmespeichern für ein Druck-luftspeicherkraftwerk". doctoral thesis. Braunschweig.

Quast, P. (1981). *The Huntorf Plant: Over 3 years operating experience with compressed air caverns*. 1. Hannover: Sonderdruck KBB.

Quast, P. and F. Crotogino (1979). "Initial Experience with the Compressed-Air Energy Storage (CAES) Project of Northwestdeutsche Kraftwerke AG (NWK) at Huntorf/West Germany: Erste Erfahrungen beim Betrieb des Luftspeicherprojektes der Nordwestdeutsche Kraftwerk AG (NWK) in Huntorf". In: *Erdoel-Erdgas-Zeitschrift* 95, pp. 310–314.

Raju, M. and S. K. Khaitan (2012). "Modeling and simulation of compressed air storage in caverns: A case study of the Huntorf plant". In: *Applied Energy* 89.1, pp. 474–481. ISSN: 03062619. DOI: 10.1016/j.apenergy.2011.08.019.

Ramadan, O., S. Omer, M. Jradi, H. Sabir, and S. Riffat (2015). "Analysis of compressed air energy storage for large-scale wind energy in Suez, Egypt". In: *International Journal of Low-Carbon Technologies*. ISSN: 1748-1317. DOI: 10.1093/ijlct/ctv007.

Rehman, S., L. M. Al-Hadhrami, and M. M. Alam (2015). "Pumped hydro energy storage system: A technological review". In: *Renewable and Sustainable Energy Reviews* 44, pp. 586–598. ISSN: 13640321. DOI: 10.1016/j.rser.2014.12.040.

Reilly, R. W. and R. B. Schainker (1982). *An Economic Comparison of CAES Hardrock Salt aquifer: Report prepared for the U.S. Department of Energy by Pacific Northwest Laboratory*. Ed. by PNL. Richland, Washington.

Rutqvist, J., H.-M. Kim, D.-W. Ryu, J.-H. Synn, and W.-K. Song (2012a). "Modeling of coupled thermodynamic and geomechanical performance of underground compressed air energy storage in lined rock caverns". In: *International Journal of Rock Mechanics and Mining Sciences* 52, pp. 71–81. ISSN: 13651609. DOI: 10.1016/j.ijrmms.2012.02.010.

— (2012b). "Modeling of coupled thermodynamic and geomechanical performance of underground compressed air energy storage in lined rock caverns". In: *International Journal of Rock Mechanics and Mining Sciences* 52, pp. 71–81. ISSN: 13651609. DOI: 10.1016/j.ijrmms.2012.02.010.

RWE Power AG, ed. (2010). *ADELE - Der adiabate Druckluftspeicher fuer die Elektrizitaetsversorgung: Project Brochure*. Essen/Köln.

Saadat, M., F. A. Shirazi, and P. Y. Li (2015). "Modeling and control of an open accumulator Compressed Air Energy Storage (CAES) system for wind turbines". In: *Applied Energy* 137, pp. 603–616. ISSN: 03062619. DOI: 10.1016/j.apenergy.2014.09.085.

Safaei Mohamadabadi, H. (2015). "Techno-Economic Assessment of the Need for Bulk Energy Storage in Low-Carbon Electricity Systems With a Focus on Compressed Air Storage (CAES)". Doctoral dissertation. Harvard: University. URL: http://nrs.harvard.edu/urn-3:HUL.InstRepos:14226038.

Schastlivtsev, A. I. and O. V. Nazarova (2016). "Hydrogen–air energy storage gas-turbine system". In: *Thermal Engineering* 63.2, pp. 107–113. ISSN: 0040-6015. DOI: 10.1134/S0040601516010109.

Scholz, R., M. Beckmann, C. Pieper, M. Muster, and R. Weber (2014). "Considerations on providing the energy needs using exclusively renewable sources: Energiewende in Germany". In: *Renewable and Sustainable Energy Reviews* 35, pp. 109–125. ISSN: 13640321. DOI: 10.1016/j.rser.2014.03.053.

Schön, J. (2015). *Physical properties of rocks: Fundamentals and principles of petrophysics*. Vol. volume 65. Developments in petroleum science. Amsterdam, Netherlands: Elsevier. ISBN: 9780081004234.

Schulz, D., ed. (2015). *Tagungsband Nachhaltige Energieversorgung und Integration von Speichern: NEIS 2015*. 1. Aufl. 2015. Wiesbaden: Springer Fachmedien Wiesbaden GmbH. ISBN: 9783658109578.

Schwoeppe, P., S. Gose, and R. Scholz (2008). "Thermodynamische Betrachtungen". In: *Grundlast von der Nordsee*. Ed. by P. Dietz. Clausthal-Zellerfeld: Papierflieger, pp. 131–146. ISBN: 978-3-89720-978-7.

Sciacovelli, A., D. Smith, H. Navarro, Y. Li, and Y. Ding (2016). "Liquid air energy storage – Operation and performance of the first pilot plant in the world". In:

Sciacovelli, A., Y. Li, H. Chen, Y. Wu, J. Wang, S. Garvey, and Y. Ding (2017). "Dynamic simulation of Adiabatic Compressed Air Energy Storage (A-CAES) plant with integrated thermal storage – Link between components performance and plant performance". In: *Applied Energy* 185, pp. 16–28. ISSN: 03062619. DOI: 10.1016/j.apenergy.2016.10.058.

Setzmann, U. and W. Wagner (1991). "A New Equation of State and Tables of Thermodynamic Properties for Methane Covering the Range from the Melting Line to 625 K at Pressures up to 100 MPa". In: *Journal of Physical and Chemical Reference Data* 20.6, pp. 1061–1155. ISSN: 0047-2689. DOI: 10.1063/1.555898.

Singh, R., R. P. Saini, and J. S. Saini (2009). "Models for Predicting Thermal Performance of Packed Bed Energy Storage System for Solar Air Heaters - A Review". In: *The Open Fuels & Energy Science Journal* 2.1, pp. 47–53. ISSN: 1876973X. DOI: 10.2174/1876973X00902010047.

Sørensen, B. (2007). *Renewable energy conversion, transmission, and storage*. Amsterdam and Boston: Elsevier / Academic Press. ISBN: 9780123742629.

Steinmann, W.-D. (2014). "The CHEST (Compressed Heat Energy STorage) concept for facility scale thermo mechanical energy storage". In: *Energy* 69, pp. 543–552. ISSN: 03605442. DOI: 10.1016/j.energy.2014.03.049.

— (2016). "Thermo Mechanical Concepts For Bulk Energy Storage: Manuscript submitted for publication, personal communication 2016". In: *(unpublished)*.

Sternfeld, H. J. (1995). "Capacity control of power stations by rocket combustor technology". In: *Acta Astronautica* 37, pp. 11–19. ISSN: 00945765. DOI: 10.1016/0094-5765(95)00074-A.

Sternfeld, H. J., O. J. Haidn, B. Potier, P. Vuillermoz, and M. Popp (1995). "International cooperation on high pressure combustion". In: *Acta Astronautica* 37, pp. 487–496. ISSN: 00945765. DOI: 10.1016/0094-5765(95)00075-B.

Stolten, D. (2010). *Hydrogen energy*. Weinheim: Wiley-VCH. ISBN: 978-3-527-32711-9.

Ströhle, J. and T. Myhrvold (2007). "An evaluation of detailed reaction mechanisms for hydrogen combustion under gas turbine conditions". In: *International Journal of Hydrogen Energy* 32.1, pp. 125–135. ISSN: 03603199. DOI: 10.1016/j.ijhydene.2006.04.005.

Succar, S. and R. H. Williams (2008). *Compressed Air Energy Storage: Theory, Resources and Applications for Wind Power: Report Energy Systems Analysis Group, Princeton University*. Ed. by Princeton University.

Sundararagavan, S. and E. Baker (2012). "Evaluating energy storage technologies for wind power integration". In: *Solar Energy* 86.9, pp. 2707–2717. ISSN: 0038092X. DOI: 10.1016/j.solener.2012.06.013.

Tada, S., H. Yoshida, R. Echigo, and Y. Oishi (1998). "A Numerical Analysis of the Air in a Cavern for the CAES-G/T: Laminar-Model Results Compared with Experiments". In: *Transactions of the Japan Society of Mechanical Engineers Series B* 64.627, pp. 3830–3837. ISSN: 0387-5016. DOI: 10.1299/kikaib.64.3830.

Tessier, M. J., M. C. Floros, L. Bouzidi, and S. S. Narine (2016). "Exergy analysis of an adiabatic compressed air energy storage system using a cascade of phase change materials". In: *Energy* 106, pp. 528–534. ISSN: 03605442. DOI: 10.1016/j.energy.2016.03.042.

Tsatsaronis, G. (1993). "Thermoeconomic analysis and optimization of energy systems". In: *Progress in Energy and Combustion Science* 19.3, pp. 227–257. ISSN: 03601285. DOI: 10.1016/0360-1285(93)90016-8.

Turner, J. A. (2004). "Sustainable Hydrogen Production: Viewpoint". In: *Science* 305, pp. 972–974. DOI: 10.1126/science.1103197.

United States Department of Energy, Electric Power Research Insitute, and Pacific Northwest Laboratory, eds. (1978-05-15). *1978 Compressed Air Energy Storage Symposium: Proceedings*. Vol. 2. Pacific Grove, California.

U.S. National Research Council, ed. (1977). *Assessment of Technology for Advanced Power Cycles: Project Report, Washington, D.C.* Washington, D.C.

VanWalleghem, C. (2015). *Hydrostor Activates World's First Utility-Scale Underwater Compressed Air Energy Storage System: Press release*. Ed. by I. Hydrostor. Toronto. URL: http://www.hydrostor.ca/.

VDI (2013). *VDI-Wärmeatlas: Mit 320 Tabellen*. 11., bearb. und erw. Aufl. Springer reference. Berlin: Springer Vieweg. ISBN: 978-3-642-19980-6.

Wang, Z., D. S.-K. Ting, R. Carriveau, W. Xiong, and Z. Wang (2016). "Design and thermodynamic analysis of a multi-level underwater compressed air energy storage system". In: *Journal of Energy Storage* 5, pp. 203–211. ISSN: 2352152X. DOI: 10.1016/j.est.2016.01.002.

Weber, G. H. and J. F. Weber (2010). *Thermodynamik der Energiesysteme: Konventionell - rationell - regenerativ*. Berlin and Offenbach: VDE-Verl. ISBN: 978-3-8007-3213-5.

Weiss, T. (2013). *Schätzung des Energiespeicherbedarfs des deutschen Elektrizitätssystems: Bericht*. Ed. by stoRE.

Weiss, T., J. Meyer, M. Plenz, and D. Schulz (2016). "Dynamische Berechnung der Stromgestehungskosten von Energiespeichern für die Energiesystemmodellierung und -einsatzplanung". In: *Zeitschrift für Energiewirtschaft* 40.1, pp. 1–14. ISSN: 0343-5377. DOI: 10.1007/s12398-015-0168-x.

Wiles, L. E. (1982). *Two-Dimensional Fluid and Thermal Analysis of Dry Porous Rock Reservoirs for CAES: Report prepared for the U.S. Department of Energy by Pacific Northwest Laboratory*. Ed. by PNL. Richland, Washington.

Wiles, L. E. and R. A. McCann (1981). *Water Coning in Porous Media for CAES: Report prepared for the U.S. Department of Energy by Pacific Northwest Laboratory*. Ed. by PNL. Richland, Washington.

Wirth, M. A. (2015-08-03). "Wirtschaftlichkeitsbetrachtung eines Druckluftspeicherkraftwerks im Residuallastbetrieb". Bachelor thesis. Dresden: TU Dresden.

Wolf, D. (2011). *Methods for Design and Application of Adiabatic Compressed Air Energy Storage Based on Dynamic Modeling*. 1st ed. Vol. 65. Umsicht- Schriftenreihe. Oberhausen: Karl Maria Laufen and Laufen. ISBN: 978-3-87468-264-0.

Wolf, D., S. Berthold, and C. Dötsch (16.06.2009). *Dynamic Simulation of possible heat management solutions for adiabatic compressed air energy storage*.

Xia, C., Y. Zhou, S. Zhou, P. Zhang, and F. Wang (2015). "A simplified and unified analytical solution for temperature and pressure variations in compressed air energy storage caverns". In: *Renewable Energy* 74, pp. 718–726. ISSN: 09601481. DOI: 10.1016/j.renene.2014.08.058.

Yang, K., Y. Zhang, X. Li, and J. Xu (2014a). "Theoretical evaluation on the impact of heat exchanger in Advanced Adiabatic Compressed Air Energy Storage system". In: *Energy Conversion and Management* 86, pp. 1031–1044. ISSN: 01968904. DOI: 10.1016/j.enconman.2014.06.062.

Yang, Z., Z. Wang, P. Ran, Z. Li, and W. Ni (2014b). "Thermodynamic analysis of a hybrid thermal-compressed air energy storage system for the integration of wind power". In: *Applied Thermal Engineering* 66.1-2, pp. 519–527. ISSN: 13594311. DOI: 10.1016/j.applthermaleng.2014.02.043.

Yao, E., H. Wang, L. Wang, G. Xi, and F. Maréchal (2016). "Thermo-economic optimization of a combined cooling, heating and power system based on small-scale compressed air energy storage". In: *Energy Conversion and Management* 118, pp. 377–386. ISSN: 01968904. DOI: 10.1016/j.enconman.2016.03.087.

York, W., M. Hughes, J. Berry, T. Russell, Y.C. Lau, Shan Liu, Michael D. Arnett, Arthur Peck, Nilesh Tralshawala, Joseph Weber, Marc Benjamin, Michelle Iduate, Jacob Kittleson, Andres Garcia-Crespo, John Delvaux, Fernando Casanova, Ben Lacy, Brian Brzek, Chris Wolfe, Pepe Palafox, Ben Ding, Bruce Badding, Dwayne McDuffie, and Christine Zemsky (2015). *Advanced IGCC/Hydrogen Gas Turbine Development: Final Technical Report*. Ed. by United States Department of Energy. New York.

Zachariah-Wolff, L. J., T. M. Egyedi, and K. Hemmes (2007). "From natural gas to hydrogen via the Wobbe index: The role of standardized gateways in sustainable infrastructure transitions". In: *International Journal of Hydrogen Energy* 32.9, pp. 1235–1245. ISSN: 03603199. DOI: 10.1016/j.ijhydene.2006.07.024.

Zaloudek, F. R. and R. W. Reilly (1982). *An Assessment of Second Generation CAES: Report prepared for the U.S. Department of Energy by Pacific Northwest Laboratory*. Ed. by PNL. Richland, Washington.

Zhang, N. and R. Cai (2002). "Analytical solutions and typical characteristics of part-load performances of single shaft gas turbine and its cogeneration". In: *Energy Conversion and Management* 43.9-12, pp. 1323–1337. ISSN: 01968904. DOI: 10.1016/S0196-8904(02)00018-3.

Zhang, Y., K. Yang, X. Li, and J. Xu (2013). "The thermodynamic effect of thermal energy storage on compressed air energy storage system". In: *Renewable Energy* 50, pp. 227–235. ISSN: 09601481. DOI: 10.1016/j.renene.2012.06.052.

Zhao, P., J. Wang, and Y. Dai (2015). "Thermodynamic analysis of an integrated energy system based on compressed air energy storage (CAES) system and Kalina cycle". In: *Energy Conversion and Management* 98, pp. 161–172. ISSN: 01968904. DOI: 10.1016/j.enconman.2015.03.094.

Zhao, P., M. Wang, J. Wang, and Y. Dai (2015). "A preliminary dynamic behaviors analysis of a hybrid energy storage system based on adiabatic compressed air energy storage and flywheel energy storage system for wind power application". In: *Energy*. ISSN: 03605442. DOI: 10.1016/j.energy.2015.03.067.

Zunft, S. (2015). *Adiabatic CAES: The ADELE-ING project: Presentation at SCCER Heat and Electricity Storage Symposium, PSI, Villigen (CH), May 5, 2015.*

Zunft, S., C. Jakiel, M. Koller, and C. Bullough (2006). *Adiabatic Compressed Air Energy Storage for the Grid Integration of Wind Power*. DOI: 10.1002/9781119197249.ch1.

Zunft, S., M. Krüger, V. Dreißigacker, P.-M. Mayer, C. Niklasch, and C. Bertsch (2012). "Adiabate Druckluftspeicher für die Elektrizitätsversogung: der ADELE Wärmespeicher". In: *Kraftwerkstechnik*. Ed. by M. Beckmann and A. Hurtado. Vol. 4. Neuruppin: TK Verlag, pp. 749–757. ISBN: 978-3-935317-87-0.

www.ingramcontent.com/pod-product-compliance
Ingram Content Group UK Ltd.
Pitfield, Milton Keynes, MK11 3LW, UK
UKHW022000190726
13853UKWH00004B/1652

9 783736 973411